TRANSPORT ECONOMICS MATTERS

TRANSPORT ECONOMICS MATTERS

Applying Economic Principles to Transportation in Great Britain

David J. Spurling, John Spurling and Mengqiu Cao

BrownWalker Press
Irvine • Boca Raton

*Transport Economics Matters: Applying Economic Principles
to Transportation in Great Britain*

Copyright © 2019 John Spurling and Mengqiu Cao.
All rights reserved. No part of this publication may be reproduced, distributed, or transmitted in any form or by any means, including photocopying, recording, or other electronic or mechanical methods, without the prior written permission of the publisher, except in the case of brief quotations embodied in critical reviews and certain other noncommercial uses permitted by copyright law.

BrownWalker Press / Universal Publishers, Inc.
Irvine • Boca Raton
USA • 2019
www.BrownWalkerPress.com
2019

ISBN: 978-1-62734-709-9 (pbk.)
ISBN: 978-1-62734-710-5 (ebk.)

Typeset by Medlar Publishing Solutions Pvt Ltd, India
Cover design by Ivan Popov

Publisher's Cataloging-in-Publication Data
provided by Five Rainbows Cataloging Services

Names: Spurling, David J., author. | Spurling, John, 1975- author. | Cao, Mengqiu, author.
Title: Transport economics matters : applying economic principles to transportation in Great Britain / David J. Spurling, John Spurling, [and] Mengqiu Cao.
Description: Irvine, CA : BrownWalker, 2019. | Includes index.
Identifiers: LCCN 2018964453 | ISBN 978-1-62734-709-9 (paperback) | ISBN 978-1-62734-710-5 (ebook)
Subjects: LCSH: Transportation. | Transportation--Costs. | Transportation--Management. | Transportation--Planning. | Transportation--Decision making. | BISAC: BUSINESS & ECONOMICS / Industries / Transportation. | BUSINESS & ECONOMICS / Strategic Planning. | TRANSPORTATION / General.
Classification: LCC HE151 .S67 2019(print) | LCC HE151 (ebook) | DDC 388/.049--dc23.

Table of Contents

About the Authors . . . . *vii*
Acknowledgements . . . . *ix*
List of Figures and Tables . . . . *xi*

Chapter 1	Introduction	1
Chapter 2	Demand for Passenger Transport	13
Chapter 3	Demand for Freight Transport	45
Chapter 4	Cost Structures	61
Chapter 5	Forms of Competition	73
Chapter 6	Pricing Policy	93
Chapter 7	Railways	107
Chapter 8	Reasons for Government Intervention	115
Chapter 9	Road Passenger Transport – Buses, Coaches, Taxis and Cars	121
Chapter 10	The Way	133
Chapter 11	Shipping, Ports and Inland Waterways	151
Chapter 12	Airports	163
Chapter 13	Aviation	185
Chapter 14	Safety	197
Chapter 15	Coordination of Transport	209
Chapter 16	The Public Sector	215
Chapter 17	Cost Benefit Analysis	235
Chapter 18	Local Transport	241
Chapter 19	Measures of Efficiency	257
Chapter 20	Transport Investment	267
Chapter 21	Economies of Scale	279

Chapter 22	Problems of the Peak. 295
Chapter 23	Rural Transport. 317
Chapter 24	The Demand for International Transport 333
Chapter 25	Developing Countries 347
Chapter 26	Recent Developments and Future of Transport by Mode. 365

Glossary of Transport Economics *387*
Index. *411*

About the Authors

David Spurling BSc, PGCE, DGA, FCILT, M.Inst.TA
David was a founder of Learning Through Cooperation Ltd and its subsidiary LTC Kenya. He wrote 21 textbooks on a range of subjects including Transport, Economics, Business, Sociology, Tourism and Accounts. He also published a book on the Sittingbourne and Kemsley Light Railway. He taught people from more than fifty countries at a range of schools and colleges. He was an Associate Professor in Transport Economics at what is now Birmingham City University.

In addition, David founded a college in Nairobi, Kenya, to provide educational assistance to a developing country, and was a member of the Nigerian Business Examinations Council. He was an examiner for the Association of Business Managers and Administrators, the Chartered Institute of Transport, and Edexcel. David was a Quaker and this influenced his views on social issues. He devised a survey on the single homeless. He regarded climate change as one of the major issues facing the world today. He was a Parliamentary candidate for Meriden and a councillor in both Essex and Kent. He was also a fellow of the Royal Statistical Society.

John Spurling BSc (Hons), DipTP, PGDip (Law), PGDip (CMI), MRTPI, MCMI
John has more than 20 years' experience of town planning within both local government and the private sector. More recently, he established his own company JM Spurling Planning Consultants Limited offering a range of town planning consultancy services.

Mengqiu Cao BSc (Hons), GDip, MSc, PhD
Mengqiu is a Senior Lecturer (Associate Professor) in Transport and Urban Planning at the University of Westminster. He was previously a Research/Teaching Assistant at the Bartlett School of Planning, University College London (UCL), and a Visiting Lecturer at the Department of Planning and Transport, University of Westminster. He has worked in academia and industry in an interdisciplinary research field, which is primarily a mixture of transport

analysis and urban planning. His current research interests include: transport planning; integrated urban planning and sustainable transport development; social equity and travel vulnerability; statistics and transport modelling; transport and climate change; freight transport and logistics; transport economics; travel behaviour, behavioural economics and well-being. In addition, he has also worked with public authorities and international funding organisations.

Acknowledgements

We would like to sincerely thank the following individuals for their contributions to the preparation of this book:

- Angela Cooper (UCL), Jessica Zhang (UCL) and Maxine Qiu (Beijing International Studies University) for their work in checking the book.
- Jeff Young and his team for typesetting and publishing the book.
- Anthea Spurling for her moral support in this endeavour.

Thanks very much also to all the people not mentioned here but who offered their help and support in the completion of this book.

List of Figures and Tables

Figure 1.1	The London congestion charge	9
Figure 2.1	Santander cycles and dockless hire bikes (e.g. Mobike) in London	18
Figure 2.2	Dockless hire bikes (e.g. Mobike and Ofo, etc.) in Beijing	19
Figure 25.1	Jeepneys – an informal public transport in the Philippines	353
Table 1.1	The number of fatalities from 2010–2017 in the UK	6
Table 2.1	Elastic demand	13
Table 2.2	Inelastic demand	14
Table 17.1	Two types of financial accounts	236
Table 20.1	An example of the payback period method of investment appraisal calculation	270
Table 20.2	An example of ARR calculation	271
Table 20.3	An example of present and future values' calculation	272
Table 20.4	An example of calculating benefits	274

CHAPTER 1

Introduction

Importance of Transport Economics

In this book, you will learn how to apply economic principles to transport. This will not only help you if you wish to study the subject for its own sake or if you work in the transport industry and wish to improve your knowledge of the subject, but will also help you if you are taking examinations. In addition, this book will also be beneficial for someone who is interested in the subject of transport economics, but has not studied economics, statistics or mathematics.

Transport is an important subject since motoring alone accounted for 14.4% of household expenditure in 2017 (ONS, 2018[1]). Of this, 36.6% was spent on purchasing vehicles, 11% on spares, accessories, repairs and servicing, 26.1% on petrol, diesel and other motor oils, 3.5% was spent on other motoring costs, 7.4% on transport fares and 15.4% on air and other travel in Great Britain (DfT, 2017[2]). The aggregated expenditure associated with motoring, such as petrol/diesel, maintenance, insurance and MOT fees, may be even higher than the cost of the vehicle itself, something which is often overlooked when purchasing a car. In car manufacturing, the aggregated costs of bulk buying materials used in the manufacturing process such as steel, account for a substantial part of the total costs, while a smaller proportion is spent on the production of high value, low bulk items such as the CPU and engine. Transport economics is therefore important since transport constitutes a major sector of the economy in most countries.

---

[1]Office for National Statistics (ONS) (2018). *Family spending in the UK: financial year ending 2017*. Available at: https://www.ons.gov.uk/peoplepopulationandcommunity/personalandhouseholdfinances/expenditure/bulletins/familyspendingintheuk/financialyearending2017 (accessed 5th August 2018).

[2]Department for Transport (DfT) (2017). *Transport expenditure*. London: DfT.

Finance and the General Economy

Like any other branch of economics, it deals with the allocation of scarce resources. These scarce resources may be vast, such as the capital required for major projects or schemes such as the Channel Tunnel, on which over £10 billion was spent before it opened in 1994. Another example is the railway tunnel through the Brenner Pass, linking Austria and Italy, which has recently reopened at a cost of several billion pounds. However, in other cases, the costs can be small, as with many road haulage firms that operate using a single van and family members as drivers and part-time bookkeepers.

Employment and the Transport Industry

Transport economists will consider the amount of labour that the transport industry uses. During the period after the regrouping of the railways in 1923 but before the Second World War, the London Midland Scottish Railway (LMS) was the largest employer in the UK. Employment patterns have changed but currently the transport sector is a major employer in the UK, particularly the road haulage sector. The wages paid may also be an important factor. Similarly, for those who work in the airline industry, the effects of strikes or stoppages on their wages can be significant. Until the 1960s and the container revolution, the ports were a major source of employment in many areas, including Tilbury in Essex, Liverpool in North West England and Glasgow on the West coast of Scotland. Whilst the ports are still important they employ far less labour as roll-on roll-off ferries tend to be used which enable vehicles to drive straight on to the ships, and containerisation, palletisation and many forms of mechanisation have all contributed to replacing the vast numbers of workers.

Land and the Transport Industry

The land required for transport use may be considerable. For example in Los Angeles, it has been estimated that about 70% of the land space is used for vehicles, both for roads and parking. In nineteenth century, the railway occupied about 30% of the land which was used for the passenger station, as well as freight lines and sidings. Landowners in the nineteenth century often resisted the efforts of the railway companies to go through their land. This altered both the location of the railways and the total costs of the original railways.

Even today, it still affects some of the variable costs associated with railways, as well as the demand for rail travel.

Allocating land for transport purposes is a major problem. Massive amounts of land can be used for additional runways or terminals at airports. It is important to consider carefully where they should be constructed, since once they are built, it is difficult to find alternative sites. Transport space may be comprised of small pieces of land, which cumulatively will add up to a considerable amount. Planning regulations usually specify that houses must have garages, although the evidence suggests that often garages are not used for their original purpose. Therefore, a large amount of land is used inefficiently. In the United Kingdom both in urban areas and suburban areas, there are often complaints about allocating land for new housing, but more land is used for the provision of roads, than for the houses themselves. In Valencia in Spain, a town with a population of about 750,000, all new garages for residential properties have to be built underground. This reduces the volume of land space needed to accommodate the car and other vehicles.

Airport Location

Modern airports are often massive and the problem of land is one which generally arouses strong feelings, although this is nothing new. The proposed third runway for Heathrow, announced in November 2007, is a case in point, sparking considerable controversy. In July 2016, Theresa May's new Conservative government postponed the announcement of a decision about which airport would have expanded capacity to cater for the expected increase in traffic. Subsequently it announced that Heathrow expansion was the preferred option and would go ahead. The Conservative government had previously appointed Lord Adonis a former Labour Cabinet minister, to head the infrastructure commission to speed up large-scale projects. After the 2017 general election, the minority Conservative government announced that it still favoured Heathrow expansion.

Port Location

The location of ports is often subject to limitation, but once the location has been determined, the sunk costs are important. Sunk costs are costs which cannot be recovered. Because of the considerable size of modern ships, most ports have moved downstream. This means that they have generally moved nearer to the sea than they were originally.

Irreversibility of Infrastructure Decisions

Transport infrastructure projects are often irreversible without major capital expenditure. However, there have been a few new ports created, such as Thames Haven in Essex, which is located on the site of a former oil refinery.

The problems of land allocation such as trying to preserve the trackbeds of railways, in case they might be used in the future or for roads, have not been satisfactorily resolved. The pricing policy for parking and roads also causes a significant number of problems, including how land is allocated for roads and whether there should be toll roads, and congestion charges.

Donald Trump and the Environment

The US president Donald Trump has vowed to abolish many environmental policies that have been introduced and has publicly stated that he does not believe in climate change, often referred to as global warming. In 2017, the US withdrew from the COP21 Paris Climate Agreement. However, not all senior members of the US Republican Party agree with Donald Trump.

Fuel and the Transport Industry

The amount of fuel required has been a problem for many non-oil producing countries, particularly since the oil price rises from Oil Petroleum Export Countries (OPEC) following the Iranian revolution in 1979. This has led many people to speculate on how far we can move away from fossil fuels. In 2015, the price of oil halved unexpectedly. Even in the USA, former president George Bush Junior had spoken about the need for Americans to reduce their dependence upon foreign oil. Barack Obama, president from 2008 to 2016, had been more receptive to the need to pay attention to climate change. There are currently many developments taking place in the use of renewable energy for transport. Lord Deben a former Conservative cabinet minister (probably better known by his original name, Selwyn Gummer) stated at a climate change conference in 2014 that whilst he was agnostic about the types of fuel used, renewable fuel prices were continuing to fall, whilst this is not true of fossil fuels.

Climate Change Convention

In September 2016, both the US and China signed the climate change convention. These two countries between them accounted for more than 40% of the total global emissions produced. The climate change convention needed 55% of countries that produce emissions to sign up in order for the agreement to be ratified. The 28 countries of the European Union also signed up to the agreement.

Social Costs

More attention has been paid to the social costs or external effects of the transport industry. The increase in the number of air journeys has caused concern and most countries signed up to the Kyoto agreement in 1997, which aimed to limit the amount of pollution, greenhouse gas emissions (GHGs), etc. produced globally.

Importance of Social Costs and Their Definition

Social costs are the total costs to the community and therefore include both private costs and external costs. External costs are costs which are not borne by the consumer or producer of goods or services. An example of this is the costs arising if a ship sinks, such as in the case of the MSC Napoli near Branscombe in Devon in January 2007. It was carrying several thousand containers and caused considerable pollution. On 13th January 2012, the luxury cruise ship Costa Concordia suffered an accident caused by the captain's reckless behaviour, which left at least 11 people on board dead, while 4,000 people had to be evacuated. The most important social cost in the UK is the 30,000 deaths from road vehicle pollution, which the cross-party environmental audit committee identified in 2010, chaired by the then Conservative MP Tim Yeo.

Social costs are particularly important in transport economics. This is because, whereas some other industries have little impact on people who are neither consumers nor producers, the same is not true of the transport industry. This is evident to economists in the case of proposals for new transport facilities such as airports or major roads or the proposed High Speed 2 link, from London to Edinburgh and Glasgow and many other major British cities in the North of England: many people have strong views about the projects even if they are not going to use the new airport, road or railway.

Other major social costs include road accidents, which in the UK account for approximately 1,500 deaths per year and about 300,000 injuries per year. This is a considerable improvement on the number of deaths and injuries that occurred in the 1970s, when about 7,000 deaths were caused each year. In general, there has been a reduction in the number of fatalities from 2010–2017 onwards (Table 1.1), given that the considerable increase in volume of traffic.

In the past, lead additives were used in petrol and caused a considerable amount of pollution. Whilst these additives have been phased out, with minor exceptions, the numbers of deaths from pollution are still very high and Ken Livingstone, the former Labour Mayor of London, cited pollution effects as one of the reasons for the expansion of the area covered by the congestion charge. In a city such as London, traffic pollution is by far the major component of pollution, since industrial pollution is relatively minor by comparison. The current Labour Mayor of London Sadiq Khan is committed to reducing pollution further, especially in the Oxford Street area, the main shopping street in London.

Table 1.1 The number of fatalities from 2010–2017 in the UK (DfT, 2018[3])

Year	Pedestrian	Pedal Cyclist	Motorcyclist Rider/ Passenger	Car Occupant	Other Road User	All Road User Groups	Percentage Change from Previous Year
2010	405	111	403	835	96	1,850	−16.7
2011	453	107	362	883	96	1,901	2.8
2012	420	118	328	801	87	1,754	−7.7
2013	398	109	331	785	90	1,713	−2.3
2014	446	113	339	797	80	1,775	3.6
2015	408	100	365	754	103	1,730	−2.5
2016	448	102	319	816	107	1,792	3.6
2017	470	101	349	787	86	1,793	0.1

The Kyoto Protocol to the United Nations Framework Convention on Climate Change

The Kyoto Protocol to the United Nations Framework Convention on Climate Change, which was signed by many countries, is an amendment to the international treaty convention on climate, assigning mandatory targets for the

---

[3]Department for Transport (DfT) (2018). *Reported road casualties – Great Britain: 2017*. London: DfT.

reduction of greenhouse gas emissions to signatory nations. In April 2016, nearly all countries signed up to the climate change convention in Paris.

Air Transport Noise

Air transport noise is difficult to avoid and any airport expansion proposals tend to provoke complaints. It is not necessarily the noise from the aircraft themselves that people object to. Ironically, part of the concerns about additional noise from any proposed expansion of Heathrow centre on the number of cars travelling to and from Heathrow, which will add to both the noise and pollution in that part of the London area.

Public Awareness

Social costs have attracted more public attention partly because of the growing awareness of climate change. The Stern Report, published in October 2006, warned of the dangers to the UK and other economies, which this could cause. In April 2016, nearly 200 countries signed up to a United Nations (UN) convention, which represents an important step in trying to reduce the level of emissions and hence global warming, more accurately called climate change.

Transport economists might also assume that as people get richer, they will become more aware of the problems of social costs, especially if they spend more money on housing. They will then be more likely to notice the noise and pollution caused by road traffic, which will prevent them from enjoying their wealth. In the UK, nearly all advertising for new housing will stress, if possible, that it is located in a quiet neighbourhood. Most new roads, apart from bypasses, trunk roads and motorways, are built so that they do not have through traffic travelling along them, and this has become increasingly popular. Aircraft noise on the other hand is more difficult to avoid since, in the London area, as is the case for many big cities, it is very difficult to find anywhere which does not suffer from this.

Interdependence of Transport

Various different modes of transport are often interdependent. Air transport usually requires people to travel to and from the airport and so the accessibility of airports has caused a number of problems, which have not always been resolved satisfactorily. There have been some newer developments, such as the railway line from Paddington to Heathrow (the Heathrow Express), which have solved some of the problems, although it could be argued that by encouraging people to go into or through London, rather than around the city, it may have added to the congestion for some other travellers instead of alleviating it. In Edinburgh in September 2007, the Scottish parliament decided to abandon the proposed airport link to Edinburgh Airport partly for this reason. There has been considerable debate about how desirable it is to make access to and from the airports easier when the world as a whole is suffering from the effects of climate change. Indeed it is often questioned whether expansion of the air travel market is desirable at all. Sometimes there is interdependence even within the same mode of transport: for example, before the 1960s many people used the railway branch lines to get to and from the intercity services.

The Beeching Report (Reshaping of British Railways) in 1963 and Privatisation from 1994

The Beeching report (Reshaping of British railways) in 1963 largely ignored accessibility to the main railway lines and focused too narrowly on the costs and revenues associated with a line, without considering the overall impact of the line on the network as a whole. It could be argued that this lack of an overarching perspective has become even more problematic since the privatisation of the railways from 1994 onwards. A book entitled "Broken Rails" by Christian Wolmar, a well-known commentator, on rail problems, makes this point very forcefully. However, the new Elizabeth line in London will make it easier for many people to be able to transfer from one railway line to another, and has the added benefit of offering a very frequent service.

Interdependence on the Road Network

The interdependence of transport can be seen in the competition for the same facilities on the road network where an increase in the number vehicles adds to the time taken for other travellers. If 10 vehicles take 10 minutes to travel along a piece of road and 11 vehicles take 11 minutes, the marginal time (i.e. additional time for travellers as a whole) is 21 minutes, and the total time will therefore be 121 minutes instead of 100 minutes. However, to the vehicle owner concerned, the decision is made on the basis of the 11 minutes that it will take them.

Because buses have to load and unload, the time they take to travel the same distance often increases over time for the same length journey and this explains why bus travel as a whole has declined in the UK. Transport economists will therefore come to the paradoxical conclusion that if the majority of travellers went by bus then everyone would spend less time travelling, but for any individual traveller, it is quicker to use the car.

There have been some exceptions to this decline in bus patronage: for example, in London congestion is so severe that fewer people as a percentage of the whole population wish to travel by car than is the case anywhere else. This lack of enthusiasm for car travel has been exacerbated by the congestion charge, which has been in force in London since February 2003. The congestion charge is now £11.50 per day for driving a vehicle within the charging zone between 07:00 and 18:00, Monday to Friday (Figure 1.1).

Figure 1.1 The London congestion charge (Photo: Mengqiu Cao).

There have also been some improvements to bus lanes and other bus priorities, which mean that the decline in bus travel times has been partly alleviated. However, in most other parts of the UK travel by bus has continued to fall.

Rationality and Transport Economics

Transport economists, like other economists, often assume that people make their decisions about whether and how to travel on rational grounds/or on the basis of marginal costs of either money or time, or a combination of these. The same is also true of freight transport. The time taken is a prime consideration. Marginal cost means the additional cost of the journey, and a great deal of transport economics is concerned with decisions made at the margin. It is untrue that people are always rational in their decision-making (e.g. behavioural economics – Professor Richard H. Thaler), since they often underestimate the costs involved. Before concluding that this is a reason for intervention, however, transport economists would need to be sure that if other stakeholders such as central or local government made decisions, these were made on a more rational basis. There has also been considerable misinformation about pollution, with many people previously assuming that diesel was preferable to petrol engines for environmental reasons, but this has since been disproved. In April 2016 Mitsubishi admitted that their figures for fuel consumption had been dishonest since at least 1992.

Transport as a Derived Demand

Transport is not usually required for its own sake, unlike, for example, the demand for the theatre or cinema tickets, but because it gives utility of place. People may wish to travel for a variety of reasons including travel to and from work, for education, for social reasons such as a desire to go to entertainment facilities, or to visit friends and relatives.

Links Between Transport Economics and Other Academic Disciplines

Transport economics is therefore often linked to other disciplines such as the built environment. It is also linked with sociology and psychology, since it is important to understand why people behave in certain ways.

Changes in Location and Hence Changes in Demand for Transport

There has been a tendency for the inner city to decline, and for what geographers refer to as the Central Business District (CBD) to remain broadly the same size while suburban areas have increased both in terms of population and area covered. Even within these broad categories, there have been significant changes. Many commercial firms have become more footloose. Footloose means that firms are not tied to any one location because of costs. This is notably the case with the service sector, since most organisations no longer need a large central office where all their files are kept. The digital revolution has meant that files can be stored on a central computer system, which is available almost anywhere. The growth of car traffic means that often firms are not tied to a central location to suit either their customers or employees. The location of industry has changed and so has the demand. Countries such as the UK, are no longer mainly industrial with a large working class, often called blue-collar workers. Therefore, the total number of people employed in manufacturing industry has fallen rapidly. The coal mining industry, which used to employ vast numbers of people (e.g. about 200,000 people in the early 1980s), currently employs very few. This has consequences for the demand for both passenger and freight transport.

Most coal miners used to live near the coal mines so that demand for passenger transport to and from work was usually low. However, demand for freight transport was heavy, and it remains important, but most coal is now imported. Commitments to reducing climate change could decrease it even further. Heavy industry, such as large steel works, has also become less important as an employer. On the other hand, light industry units, such as those making plastics, have expanded. These are often found on industrial sites away from town centres and are often not very well served by public transport, especially rail.

Importance of Travel Surveys

Travel surveys carried out by the government (e.g. London Travel Demand Survey or UK National Travel Survey) or private organisations will help to identity these broad trends but, for transport operators as well as the providers of these facilities, smaller local surveys may be helpful. Local authorities will estimate footfall (i.e. the number of people passing a particular location).

For instance, in the UK, Passenger Focus which was renamed Transport Focus in March 2015, carries out many surveys for the government.

Changes in Shopping Patterns

Retailers or shop-owners will want to know about footfall before deciding on a location. Shopping centres have become very large in many cases (e.g. Lakeside in Essex and Bluewater in Kent, as well as the Metro centre in Gateshead in the Tyne region which has over 300 retailers and leisure facilities). This in turn generates traffic from both potential customers as well as people working in these centres. It also means that the pattern of freight demand is altered.

Lack of Homogeneous Demand for Transport

In the case of air transport, and sometimes railway travel, there are often many different fares for the same journey.

The prices that cruise ships charge also often differ significantly between those for a luxury cabin and at the other end of the scale for a room in dormitory class.

Total Journey Time and Importance in Decision-Making

Transport economists have observed that whether one travels in a mini or the latest limousine in urban areas, the journey often takes the same length of time. Sometimes the journey may even be quicker in a smaller vehicle, since it will be easier to find a suitable parking space. As far back as 1966, a House of Lords committee stated that small electric vehicles should have priority in towns.

Social Media and Information

One of the criticisms which many passengers make is that when travel disruptions occur, operators do not use social media enough to inform them about what is happening. This was particularly apparent when Heathrow and Gatwick British Airways services were disrupted on 27th May 2017 by a computer glitch. Passengers complained that no staff were available at the desks at both airports to answer any questions.

CHAPTER 2

Demand for Passenger Transport

In general, most economists use the phrase 'market demand curve' to show how much a particular good or service consumers in the market will wish to buy over a range of prices at a particular time. The market demand curve is the sum of the individual demand curves showing the demand for a particular service such as transport by individual consumers. The demand is always over a particular period. It is meaningless to say that there is a demand for 1,000 journeys unless economists mention the period (e.g. per week or year). Economists use the phrase 'elasticity of demand' to show how one variable affects the demand for a service such as transport. The most common form of elasticity of demand used by economists is price elasticity. The formula for the price elasticity of demand approximates to:

$$\frac{\text{Percentage change in quantity demanded}}{\text{Percentage change in price}}$$

Transport economists would normally expect demand to fall if price rises and conversely to rise if price falls, and therefore to have a downward sloping demand curve. Thus, the price elasticity is negative, but economists often disregard the minus sign. Economists would state that demand is elastic if total expenditure rises as price falls or if total expenditure falls as price rises, whilst demand is said to be inelastic if total expenditure rises as price rises. If economists estimate part of the following demand schedules as follows:

Then in Table 2.1, the demand is elastic.

Table 2.1 Elastic demand

Fares	Demand	Revenue
£10.00	100	£1,000
£9.50	120	£1,140

Then in Table 2.2, the demand is inelastic.

Table 2.2 Inelastic demand

Fares	Demand	Revenue
£10.00	100	£1,000
£9.50	101	£959.50

Why Are Some Demands Likely to Be Elastic and Others Inelastic?

Generally, transport economists would expect the demand for business and educational traffic to be more inelastic than demand for social travel, although there may be some exceptions to this rule. This is because often in the short run, there is no substitute for travelling to and from work, schools or colleges, although there may be a choice of modes of transport, which the customer may use. This is one reason why peak fares are often higher in the morning or evening rush-hours than at other times of the day. A price reduction is unlikely to lead to a more than proportionate increase in demand and often the mode of transport is operating at full capacity, so it could not cope with additional demand without considerable expense. Transport economists can observe this in many rail commuter services coming into London and other major conurbations in different countries. The majority of peak traffic in conurbations is usually commuter traffic and traffic to and from educational establishments play a role.

Short and Long Run Elasticities of Demand

Transport economists will probably find that even with short run elasticities of demand, a fare change upwards of 20% will have a different effect to that of a fare change downwards. This is particularly likely to be true of bus journeys, since often the existing passengers may well have alternatives including the use of the car, which can be used as a substitute quickly. However, even if fares going down have an effect on existing passengers to encourage them to make more or longer journeys, the effect on people who do not use the bus regularly may well take more time to filter through. In the longer term, it may affect the location of either work or housing. Therefore, often the elasticity figures are likely to be higher in the long rather than the short term.

The Nature of Educational Demand

The nature of educational demand is changing. In the 1960s, most students in the UK used to travel quite long distances to and from university and few people went to a local university. In the early 1960s, about 2% of the total relevant age group went to university. Now because of the absence of grants and the likelihood of tuition fees, more people may wish to go to a local university. More people go to university and the previous Labour governments from 1997 until 2010 had set a target that 50% of the relevant age groups should go to university or some other form of higher education. Whilst the present Conservative government has not set such a target, the number of people going to university or other forms of higher education has still increased. On the other hand, in the UK, there are fewer people under 16 than over 65. Fewer people will be at school in spite of the raising of the age limit to 16 in 1972. The current government has recently changed the law so that children starting secondary schools in England will be legally required to stay in education until they are 17 and this was increased to 18 for school leavers who started at school from 2005. This will affect future total demand for educational travel.

Road Pricing

The idea of charging more in peak times has been important in different modes of transport, whether rail, bus or coach. There are often seasonal patterns for air travel but not so much for roads except for tourist purposes.

Road pricing, however, has not applied generally to road traffic, except in Singapore although the London area introduces a congestion charge from 2003 and the area was extended in February 2007. There is also the M6 toll, where motorists can choose to pay to use the 27 mile route. The toll is higher during the day (06:00–23:00) than at night (23:00–06:00). Transport economists would state that the price charged should relate to the marginal social cost of congestion and pollution. This would help to give the optimum utilisation of road space.

Social Travel

For social travel, there are often a large number of substitutes including sometimes not travelling at all. Mobile phones, texting, social media, WhatsApp,

WeChat and Skype may also be a substitute for personal travel. It was often assumed that people would spend more time with home entertainment including television, computers and computer games (e.g. including virtual reality and/or augmented reality) so that demand for travel to and from cinema would inevitably decline, although in the last few years in the UK this has been less true. The decline in the number of cinemas and hence transport demand has been partially offset by the growth in the number of Bingo halls often using the old cinema sites. In China, a growing number of the younger generation are going less to the cinema instead of watching movies online at home, unless they may socialise with their friends or perhaps for dating.

Demand for Travel to and from Entertainment

There are now many multiplex cinemas and these have generated considerable demand for travel to and from them. Families often live further apart than they did in the past and surveys by Wilmott and Young in their books have shown how patterns of family change affect travel to see other family members. The publication "Social trends" now entitled "Measure of well-being" gives an indication of how frequently people see other members of their family.

Demand for Travel for Health Care

The demand for health care can generate considerable amounts of traffic, especially when cottage hospitals (i.e. smaller hospitals were often closed and larger ones opened instead). Hospital patients are often elderly and therefore less likely to be car owners. Visiting patients may pose problems especially for non-car owners, as newer hospitals are less likely to be in town centres (e.g. the new District General Hospital constructed on the outskirts of Tunbridge Wells). Hospitals have come under media pressure for charging for car parks. Whilst few people would wish to charge high prices for people visiting a dying patent, there is no reason why hospital car parks should be subsidised universally. Less media scrutiny has been made about the problem of sparse public transport services to many hospitals and the high public transport fares, which might be charged.

Income Elasticity of Demand

Economists also use the phrase income elasticity of demand to measure the responsiveness of demand to changes in income. The income elasticity of demand is positive for normal goods and services and negative for inferior goods and services. Inferior is a technical term and does not refer to the quality of the service. It has often been assumed that cycling is an example of an inferior good. This means that as incomes rise, people will turn to other modes of transport. If a country has higher incomes, it has less cycling. The assumption is that in a developed country, that demand for car transport would be regarded as a normal good or service whilst cycling would be an inferior good or service. However, this would not necessarily be true in all developed or developing countries since in the Netherlands, Scandinavia and Germany, cycling is regarded as an ordinary activity and in these countries, it is possible to see many commuters travelling by bike. Indeed, in Germany, although car ownership is much higher proportionately than in the UK, the bike share of trips in Germany is almost ten times higher than in the UK according to John Pucher and Ralph Buehler. Their article Cycling for Everyone: lessons from Europe was published in the Journal of the Transportation Research Board.

Demand for Cycling

The British government hoped that following British success in cycling events in the Olympics and Paralympics 2012 and 2016 that more people will take up cycling in the UK. In the London area, it has become easier to hire bicycles (e.g. Santander cycles, Mobike or Ofo – Figure 2.1) at many railway and Underground stations for a relatively small charge. Yorkshire Bank Bike Library was shown in a documentary by Ben Fogel that people can borrow bikes without charge and many do this even on the severe slopes of parts of Yorkshire as well as easier cycle rides for example along the canal towpaths from Hebden Bridge.

Demand for Cycling in China

On the other hand, in China, cycling has been a conventional form of transport for many people and with pollution levels very high in many cities, concern has been expressed about the increase in pollution which will occur if unconstrained car use in line with the great economic growth continues. However, it could be a special case, if there is a niche market coming out suddenly. For instance, a socio-technical transition (see a research conducted by Professor Frank Geels) may lead to a 'revolution' and it therefore encourages more people to cycle – a country such as China (e.g. Mobike and Ofo – Figure 2.2).

Figure 2.1 Santander cycles and dockless hire bikes (e.g. Mobike) in London (Photo: Mengqiu Cao).

Income Elasticity of Demand

Income elasticity of demand is approximated to by the following formula:

$$\frac{\text{Percentage change in quantity demanded}}{\text{Percentage change in income}}$$

Organisations will usually try to enter markets for normal goods or services, if they assume that incomes will rise. Sometimes this may not be possible, as a road passenger operator may not be able to move into another mode of transport. However, the road passenger undertaking may be able to offer better quality on long distance routes being able to offer food or seating that is more comfortable or provide even better toilet provision. On older cruise ships, there are moves to have slightly fewer but better class cabins.

Importance of Inferior and Normal Goods and Services

Sometimes the transport organisation may try to alter the image of their product so that it becomes a normal good. In the UK, British Rail had regarded

Figure 2.2 Dockless hire bikes (e.g. Mobike and Ofo, etc.) in Beijing (Photo: Mengqiu Cao).

steam as old-fashioned. It had phased out all steam services by 1968 with the exception of the narrow-gauge line from Aberystwyth to Devils Bridge in Mid Wales. In recent years, some operators, such as the Cathedrals Express, have found that there is a niche market for longer distance steam hauled trains to and from the cathedral cities. As the name implies, the term Cathedrals Express is a slightly misleading title as it does now offer trips to some other destinations as well. Many voluntary railways sometimes known as heritage railways have also found that there is a niche for steam hauled services although often this is supplemented by old-fashioned stations.

Use of Cross-Sectional Data

Cross-sectional data means looking across different sections of society, often through looking at different income brackets for households at the same point of time or without regard to differences in time. Economists would normally need to modify this slightly since clearly a single person with a household income of £30,000 will have a different pattern of demand to a family with four children with the same total household income. In practice, economists would need to carry out surveys to try to find out whose household incomes are in this amount, since whilst Income Tax returns will give an indication (assuming no avoidance or evasion) of individuals' incomes, it does not give us an indication of the total household income. The 2011 national census in the UK is a useful source of information.

Transport economists can often use cross-sectional data to find out what happens with different income groups to demand for goods and services including public transport. This is then often used when forecasting future demand. However, care has to be taken when doing this. If economists look at people whose household income is around £30,000 per year, economists cannot assume what people whose income is currently around £20,000 will do when their income rises to £30,000, but in real terms have the same pattern of demand. It might give realistic forecasts for car ownership but be less accurate for leisure travel.

Problems of Obtaining Cross-Sectional Data

There may be problems in distinguishing between different data. This could include different income groups and transport economists will wish to know whether different travel patterns result from socio economic class rather than income. Some of these may be obvious. People in higher social groups are much more likely to visit the opera house, theatre or ballet than lower income groups. Whilst the stereotype football supporters would be older working-class males, supporters of the main football clubs such as Chelsea perhaps because of the high price of tickets are much more likely to be well off earners, often in their 30s or 40s. If transport organisations are catering for demand to and from sporting events, better knowledge of socio-economic class will be important when trying to forecast demand by different modes of transport.

Time Series Data

Transport economists often use time series data such as those found in Transport Statistics (an annual publication, which was issued by the Department for Transport). It is now possible to obtain considerable amounts of transport data free of charge from government websites. If, however, transport economists are using them for forecasting, it is sometimes better to look at disaggregated data rather than the aggregated data, since in the longer terms, some trends might not be likely to continue which would affect our forecasts.

If transport economists were doing a long-term forecast, they might assume that for example, the number of trips made by women would increase to the point where there were similar numbers of women and men drivers. This may occur, as the younger generation are much more likely to have this pattern than the traditional pattern of male drivers and women passengers.

Aggregated data may also conceal a number of different trends, which will be of interest to both transport operators as well as to the government. Total numbers of air passengers are important, but more important for the provision of airports will be where the passengers have their origins and destinations. Transport economists call this the origin/destination matrix. If the majority of air travel were from people who wanted to cross the Atlantic, this would mean that there would be a need for airports to the West of the UK. However, if the majority of passengers wish to go to and from the Continent especially Spain or Italy, it would probably mean that the United Kingdom needs greater capacity to deal with this in the South East of England.

Geometric or Arithmetic Growth Patterns

Geometric growth would mean that there was the same percentage of growth in each year. In Communist China, the Gross Domestic Product (GDP) has been increasing by 7%–10% per year, so starting from a base of 100, then in the first year, it will rise to 110 (e.g. 100 increase by 10%), but in the second year to 121, since there is 10% growth on this additional 10%. A geometric progression would mean that it would be 100, 110, 120, etc.

In the short term if the UK has economic growth of around 2%–3% which has been customary since the Second World War, then there is likely to be similar growth in transport demand. Since the credit crunch 2007 onwards, growth rates have fallen, not just in the UK, but also throughout most Western and Eastern Europe. In the longer term, however, they would give very different results. Whereas the Rev Thomas Malthus (1766–1834) a well-known eighteenth century economist gave the very gloomy prediction that food would increase in arithmetic progression and population in a geometric progression, ordinary people have become used to thinking of growth in geometric terms. How far climate change will affect future growth is a matter of debate although not that it will do so.

Limits of Traffic Growth

It is difficult to say what these limits will be. For most people, there will be a maximum time, they would choose to travel, but this will be determined partly by other factors, such as time involved in travelling to and from employment or caring for dependents. In practice, transport economists can assume that few people would wish to spend most of their time travelling, although these limits may be partially determined by geographical factors in the country. For car drivers, it is easier to drive for long periods on roads, with few intersections, such as the Pacific Highway in Australia, than it would be to drive equivalent distances in the UK. There is a steady rise in the amount of time that people spend commuting in the UK. Shocks to the system could involve congestion charging being used in more towns, which might mean more car sharing or possibly mere transfer of timings of journeys. One of the practical problems is that measures, intended to reduce problems may have the side effect of adding to problems. This is sometimes known as the law of unintentional consequences. Measures to reduce congestion, may add to the problem in the longer term. The Leitch committee 1979 concluded that road building which aimed to reduce the problems often generated more traffic.

Cross Elasticity of Demand

Economists also use the phrase cross elasticity of demand. The formula for the cross elasticity of demand is approximately:

$$\frac{\text{Percentage change in quantity demanded of X}}{\text{Percentage change in the price of Y}}$$

If the price of petrol went up by 10%, then economists might expect demand for car travel to fall by 3%, so the figure quoted would be – 0.3. Perhaps more importantly if an increase in the price of travel by car was to rise significantly (e.g. perhaps because of road congestion pricing), but little happens to demand for rail or road passenger transport, then the private vehicle is in effect a monopoly provider.

Where services are complementary as with a railway branch line acting as a feeder service for the main line the two services are complementary. If the price of the feeder service to the station rises, then the demand for the main line service will fall. The complementary good or service is not necessarily the same mode of transport and rail operators have been concerned about the low price of parking at some stations. This means that, given the opportunity cost of land, non-motorists are cross subsiding car owners. On the other hand, if parking charges get raised too much, rail operators will wish to determine that they do not lose too many of the rail passengers. There has been a concern about the impact of large car parks at the station at Ebbsfleet International in Kent on the Channel Tunnel route. If the demand is high, this will create more problems of congestion and pollution around the area. An example of this occurred when Eurostar in 2009 was busy advertising how much less pollution the Eurostar has compared with flying on the same journeys.

The Meaning of the Value of Cross Elasticity

The higher the negative value of the cross elasticity of demand the more complementary are the two services. From the government's viewpoint, one of the important factors when looking at forecasting demand for transport is the relationship between fuel consumption and the demand for road traffic. Transport economists would generally assume that if the price of petrol rose rapidly then the demand for car travel, compared to the demand that would have prevailed if fuel price had remained at the same price in real terms. Similarly, the halving in oil prices in 2015 will have led to more demand for road transport. Real terms mean after allowing for inflation. Economists also use cross price elasticity of demand to measure monopoly power.

If changes in the price of rail fares do not lead to any great change of demand for bus transport for the same journeys, economists will assume that the rail operator has a degree of monopoly power. Some economists do not accept the concept of rationality. Often potential passengers may not be aware of changes in fares for different modes of transport. This is partly because passengers do not have perfect information. Poor marketing is often a neglected factor in transport economics but many transport companies now including the rail operators are using social media to keep people informed of what is happening.

Cross Elasticity and Substitution

If the rise in the price of X leads to a rise in the demand for Y (i.e. if there is positive cross elasticity), then these two goods are said to be substitutes. The 1985 Transport Act, led to more competition for bus services. This meant that in a town, such as Harrogate in Yorkshire, almost immediately after the 1985 Act the two rival bus services when going between two centres such as Harrogate and Knaresborough about 10 km apart, were close substitutes. The higher the positive value of the cross elasticity of demand the nearer the two goods or services are to being perfect substitutes. The concept of a perfect substitute is even more applicable in freight transport than in passenger transport where one tramp ship operator's services will be regarded as an almost perfect substitute for another. Tramp here does not mean a poor service but simply one which one, which reacts to demand rather than a timetabled service.

Importance of Cross Price Elasticity

Both central government and local government in the UK have held down fares partly in order to reduce social costs, such as congestion. If, however, reducing the prices below that which would be charged in marginal cost or even average cost pricing then a low cross price elasticity would suggest that cheaper fares are not likely to be a very effective method of doing this. Transport economists would also need to know what the elasticity figures are in both the peak and off peak. If they obtain figures which show that the figures are about 0.3 or 0.4, they would come to the conclusion that fare changes on their own are unlikely to be effective. They would, however, need to go further to see why it does not have much effect. If the main reason is that motorists are often unaware of what public transport fares are then better publicity methods might be a necessary condition in order to gain more customers. Bus operators have not always tried hard enough to make people aware of their fares.

Perceived Costs and Demand

Apart from the influences of fares on the passenger service and its substitutes and the incomes of the individuals using the service, a number of other factors may also influence the demand for passenger transport. Sometimes the perceived costs of transport may be a more important influence than the actual price. It is highly likely that some costs, such as petrol and diesel, are very noticeable whereas other costs, such as depreciation and maintenance, are often drastically underestimated. This distorts the market. In the passenger transport sector, there is considerable evidence to suggest that motorists are not fully aware of their costs. The main perceived costs include fuel, both petrol and oil and parking. Motorists are much less likely to be aware of total maintenance costs over the lifetime of the vehicle or even of depreciation. It is therefore sensible to look at the index for petrol and oil and to compare it with the retail price index.

Using Published Data to Look at Comparative Costs of Motoring and Public Transport

Readers can look at Transport Trends for the latest information about real costs of motoring including the cost of fuel and compare this with public transport fares for both rail and bus. In January 2018, there was considerable publicity about the ways that railway fares were being increased on average at their highest level for 5 years.

Effect of Perceived Costs

If motorists do underestimate their costs, then their demand will be higher than would apply if they acted on perfect information. Public transport operators will therefore have to take into account the motorists' perceived costs and try to offer the same type of pricing policy (e.g. perhaps with the issue of some form of travel cards, which then gives cheaper rates once this amount has been paid). There are a number of different pricing policies, which can be pursued. The government may also need to look at perceived costs when formulating its investment policy, including road. If the government really wished to reduce demand for car travel, then it might be sensible to raise fuel taxes even if this is unpopular. The UK government did try to bring in an automatic fuel tax increase, although partly because of fuel protests in 2001 this has not always been pursued.

Quality of Service and Demand

The quality of service, particularly frequency and time of journey may be a very important factor particularly for wealthier individuals. Transport economists can observe this with taxis, which are more likely in Central London to be used by the wealthy rather than by the poor. Similarly, the time taken for the overall journey will be important. The total time taken for a journey can be subdivided into three elements, namely accessibility time, waiting time and journey time. It is possible to have a formula such as:

$$\frac{\% \text{ change in demand}}{\% \text{ change in overall time}}$$

Economists would expect with minor exceptions such as cruise ships to find a negative correlation. Therefore, rail operators will hope that a speeding up of rail journeys on the West Coast main line from London to Glasgow will have a positive effect on the demand for travel.

However, transport economists have found that overall journey time whilst important is not a good enough measure. People value a reduction in accessibility time or waiting time as usually more important than a reduction in the journey time of the main stage of transport. In turn, transport operators have realised this and sometimes have tried to ensure that the waiting time is less unpleasant than waiting at a draughty bus stop or railway station without an adequate shelter. Airline passengers typically have to wait the longest, especially since the Twin Towers tragedy in 2001. Terrorist attacks in 2015, in Paris and Brussels and in Nice in 2016, Manchester 2017 will also have added to security problems, and therefore to waiting times. Airports will often because of long waiting times offer shopping facilities or places to eat, which reduce the impact of this waiting time. The withdrawal of Britain from the EU may also cause longer waiting times for British people visiting other European Union countries and also for EU passengers visiting the UK.

Demographic Aspects and Demand

Demography including the size of family and age of people using transport services is important. In the United Kingdom, there is an aging population with more people over 65 than under 16. The retirement age conventions are important, as generally the demand for travel to and from work will usually be much smaller for pensioners than for men in the 30–40 bracket. John Cridland CBE was appointed in 2016 to recommend what the state retirement age should be for both men and women. The Turner commission in November 2005 recommended that the retirement age for both men and women should be equal and that it should be gradually raised. A report from the House of Commons in March 2016 also published proposals about this, Sometimes the differences in employment patterns between men and women, the gender gap, is important in predicting demand. This, however, will vary across countries, since in Scandinavia, a higher proportion of women will work than in the UK. Transport economists can observe from government publications such as the measure of well-being that women are less likely to own cars and to be less likely to have a driving licence, although the gender gap is much bigger with the older rather than with the younger age groups. Transport economists will be interested in the data about male and female car ownership.

The Importance of Habit

Whilst economists have often assumed, particularly in the model of perfect competition, that consumers are rational and weigh up all the options, this is not necessarily true. Trying to get information may be time consuming and difficult. Whilst the information service about the railways is supposed to offer impartial advice, it has not always been able to offer accurate information on the wide range of fares, which are now available across different operators. Similarly, bus operators as a whole have not offered a readily available source of information on bus services in different parts of the country. The internet has made it easier to obtain information using websites such as "Transport Direct". Even with airlines where because of generally higher fares, it is easier to spend money on marketing more cost effectively, information about fares has not always been easy to obtain, although the internet has often made the market more competitive.

Habit and Demand

Habit is important to many people such as many people in London who will probably use the Underground services (more commonly referred to as the Tube), even if this is not necessarily the quickest or cheapest form of transport. Car drivers often will not be aware of public transport services and will not therefore use them unless they are drawn to their attention.

Land Use Planning

Land use planning or its absence may also be important. The Buchanan report on "Traffic in Towns" published in 1963 looked at the effect of transport and town planning on demand. Currently, local authorities are looking at the effect of land use planning on sustainable growth. This is because the location of housing, schools, hospitals and work places, will have an impact on both the total demand and the modal spilt. It recognises that for too long politicians and other stakeholders have looked at mobility rather than accessibility.

How Can Transport Economists Measure Passenger Transport Demand?

Transport economists can measure demand for passenger transport in a number of different ways. These include passenger kilometres, vehicle kilometres or passenger journeys. Which of these measures is appropriate will depend upon the aims of using the information. A transport planner or highways engineer wishing to determine the capacity of a road system will be interested in the number of vehicles and types of vehicles operating at a given time especially in the peak periods rather than in the total number of people travelling in those vehicles. If, however, people were to share cars then the same number of people could be accepted on to the road network. In the USA, sometimes priority is given to cars with two or more people. One of the side effects of a congestion charge as in London may be to encourage some regular commuters to share cars, which would be helpful in reducing congestion. In this case, the number of people coming into towns by car could stay constant but if there were fewer cars, this would be useful in reducing both congestion as well as pollution.

Why Different Groups Would Wish to Forecast Future Demands for Transport?

The highway engineer when considering road maintenance will be interested mainly in the heavier vehicles rather than the lighter vehicles and in particular those vehicles, which have a high weight imposed per axle. When looking at capacity the total number of vehicles will be important. If vehicles are moving at different speeds, the capacity of the road system will be lower than if they are travelling at more even speeds. The higher the speeds generally the lower the capacity since the headway for vehicles should be greater. The passenger transport operator will be interested in the number of vehicle kilometres when considering the costs of providing the passenger service. The commercial side of a passenger operation may be interested in the mixture of passenger kilometres and passenger journeys when looking at the potential revenue of a new service.

The Department for Business, Enterprise and Regulatory Reform (which is responsible for energy) may be interested in the number of vehicle miles by different types of vehicles and their method of traction (e.g. whether they use derv (diesel), petrol or electricity battery-operated milk floats more recently hybrid and electric cars). They may also be interested in the potential for cars using hydrogen cells, which could have many advantages.

They will be interested in the degree of passenger kilometres travelled on electrified and non-electrified railway lines, both now and in the future. Economists have often assumed that it would be better to use electrified lines, since this would be helpful to reduce demand of fossil fuels. In the UK, according to the Department for Transport Publication Transport Trends Statistics Great Britain, the number of passenger kilometres has increased steadily. The main increase has been in terms of travel by cars and taxis.

The Labour governments which were in power from 1997–2010 had originally assumed that they could reduce the volume of car traffic, although they were not very successful. The coalition government 2010 to 2015 and the current Conservative government have plans to electrify lines from London to Cardiff to Swansea, and the Valley lines around Cardiff. The electrification of the line from London to Oxford will also help.

The volume of rail transport has decreased in relative terms for passenger transport, compared with other modes of transport particularly the car.

Rail passenger traffic has doubled since privatisation in 1994. It is difficult to determine any one factor. In the boom period in the late 1980s, rail passenger transport also increased, when British Rail was in the public sector. There have been quite different trends in demand between different companies even allowing for the differences in types of travel for the different companies. Some companies such as the former Midland Main Line operating from St. Pancras have been very successful in getting new passengers as has also the operator from Marylebone Station (Chiltern Railways).

Problems of Obtaining Data

Often data are not readily available. In the case of private transport such as cars, motor bikes and cycles there is no automatic source of data so that surveys will have to be carried out whether on motorways or other types of roads. It is easy to count the number of vehicles at any one point and the types of vehicles but this does not help to indicate ultimate destination or origin. Partly for this reason, therefore Oxford University introduced a Household Activity Travel Survey (HATS) system which looked at the demand for cars, etc. by looking at the total demand during the day. Often a mother might take the children to school in a car and then go onto either full time or part time work in her car. This type of analysis is useful since if the aim is to reduce car dependence, the purpose of car journeys need to be known. For public transport, data are more readily available on the number of passenger journeys rather than for passenger kilometres. However, increasingly in conurbations there have been season tickets, travel cards and free bus fares for many old age pensioners which means that data are more difficult to come by. Even where tickets are issued for each journey if there is a coarse fare structure, for example a fare of £2.00 for 7–10 kilometres for the West Midlands in the 1970s all figures will be estimates unless surveys were carried out by the West Midlands Passenger Transport Executive. Where there is a flat fare system or a zonal system the figures for passenger kilometres will be an estimate unless, as in the West Midlands, field surveys are carried out. Even the number of passenger journeys may not be wholly reliable if a number of different tickets are given for one journey. Currently there has been some criticism that some of the apparent increase in the number of rail journeys has been that because there were several operators, in some cases passengers had to obtain two tickets, whereas previously they would have had one from British Rail. Because of the fare systems, it was sometimes cheaper to buy two tickets rather than one. The trend, however, towards computerised ticket issuing should enable the operator to have a more comprehensive idea of where passengers are boarding though not necessarily, where they are alighting. With the London Underground, data can be available since passengers have to go through barriers at both ends.

One of the criticisms of data in the UK that until relatively recently data for walking or cycling were often virtually non-existent. The National Travel Survey in 1976 to 1977 provided more data. Since that time, the government has said that it wants to encourage cycling partly for environmental reasons and partly because cycling is a healthy activity. However, the whole amount spent on the Sustrans national network is only £200 million and even the funding available for the existing six demonstration towns as well as new cycling towns is less than £100 million. These figures are very small compared with even a modest road improvement scheme. Sustrans stands for (Sustainable Transport) and the cycle network extends from Inverness in North Scotland to Dover in Kent. Following a poll by the Peoples national lottery, an extra £50 million has been made available to Sustrans, where it was the winner from four applicants. However, this is an odd way of funding transport projects.

Cross Elasticity of Demand

The cross elasticity of demand between different modes and within modes of transport is important but is often difficult to find out. This is because often a whole variety of different factors are at work, changes in prices of cars, petrol or diesel, different types of public transport as well as changes in location.

There is likely to be a difference between the short run cross elasticity and the long run cross elasticity. If bus or rail fares rise considerably at one time, whilst the cost of running a car increases gradually, there may not be an immediate transfer from bus or rail to car but in the longer-term economists might expect people to buy more cars. Generally, economists would therefore expect the short run price elasticity to be lower than the long run.

Conversely, fare rises in the real or perceived cost in motoring or a reduction in fares may not immediately persuade people to switch from cars to public transport. In the longer term, if fares were held down for a long period as within the South Yorkshire Passenger Transport Executive in the 1970s, economists might expect younger people particularly in the 20–25 age group to have deferred purchases of cars.

Income Elasticity of Demand

For most goods and services as incomes rise, demand also rises. Transport is no exception to this. However, whilst overall expenditure on transport has remained reasonably constant this hides a number of different trends. Air transport almost certainly has an income elasticity of demand, which is greater than one. In the early 2000s, there was considerable debate about how far government should allow demand to rise by building more runways or allowing the expansion of airports in other ways. Critics suggest that we are not comparing modes in a similar way, since the aviation industry does not pay the same rates of fuel duty as other modes of transport and currently air transport is one of the rapidly growing sources of pollution. In 2014, UK residents made 60.1 million visits overall, which was almost double the 34.4 million visits to the UK by overseas residents.

80% of UK residents' overseas travel was by air, 12% by sea and 8% by the channel tunnel. Transport economists would wish to distinguish between demand, which occurs because of market forces and those which occur because of either taxation or subsidies. The advent of budget airlines and the increased use of the internet to be able to find bargains in travel have also contributed to the increase in demand for air travel.

Whilst there has been a considerable degree of analysis of subsidies paid to public transport, the subsidies paid to company cars have been subject to less analysis. A 1979 British Institute of Management survey stated that about 70% of new cars were subsidised by companies in some way. The then Chancellor of the Exchequer in 1988, Nigel Lawson (now Lord Lawson) announced that he was trying to phase out the subsidies to company cars and doubled the amount of tax which was payable on most of them. He also announced that he would have liked to have eliminated the subsidies paid to company car parking, but could find no way of doing this. Some press reports in 1988 stated that in London the cost of a car parking space was £4,000 per place. Planning laws could be used to reduce demand by insisting on a maximum number of parking spaces to be provided rather than a minimum and currently the government is doing this. The payments made for vehicle excise duty (VED) have been changed in recent years partly to try to persuade people to take more notice of fuel consumption and emissions and to persuade them to use fewer "gas guzzling" vehicles to use the colloquial jargon.

Location Policy and Transport Demand

As transport is a derived demand, the location of offices such as moving away from Central London will affect demand. The regional policy in the 1960s and the early part of the 1970s encouraged moves away from overcrowded centres particularly London which had the effect of reducing journey time and distance to and from work, but perhaps increasing inter-city travel by businessmen. The effect of the new towns such as Milton Keynes, Basildon is difficult to determine. In the early part of the twenty-first century, the government announced plans to develop the Thames Gateway area with many new houses being built on both sides of the River Thames. In 2007, the then Labour government announced that it would be the first eco-region.

The Thames Gateway plan looks at past data, partly to try to forecast future travel patterns. This is partly complicated by the fact that in early 2007 there were 10 different train operating companies, including Chiltern, Silver link, Virgin Cross Country, Virgin West Coast, First Great Western, South West Trains, Southern, South East, Heathrow Express and Gatwick Express. Currently there is an analysis of the past using historic ticket sales.

Changes in factory patterns from the older areas along railway lines, canals and rivers to the outskirts of towns also affect both total demand and modal split. In general, public transport finds it easier to cater for a radial pattern rather than as in Coventry factories in a circular pattern. This is particularly true of tracked systems whether rail or trams or trolleybuses. The tendency from about 1977 onwards to have more emphasis upon the inner-city areas accentuated by the Toxteth riots in 1981, may also affect demand. The 2011 riots in England and Wales, but not Scotland will have emphasised the need for action to take place to prevent future riots. The changes made by the former London Docklands Development Corporation (LDDC) 1981 to 1998, which was both a large-scale landlord but which also had powers to alter the transport system meant that many people were able to live nearer their work and therefore that total transport fell. The extension of the Jubilee line to Stratford as well as the extension of the Docklands Light Railway has dramatically reduced the demand for car travel in that area. The overground, which passes through Stratford in East London, has also been successful. Crossrail which has been renamed the Elizabeth line, will launch at the end of 2020 when the central tunnels under London open, and will make crossing London travel easier. It will be a very frequent service, which means that waiting time will be very small.

The urban/rural/suburban split may also affect both the total demand for transport and the modal split. Car ownership is much higher in rural areas than in suburban areas and people in urban areas have lower car ownership ratios even for the same level of incomes.

Substitutes for Transport

The telephone service has always been a substitute for postal services but also acts as a substitute for passenger transport, since often a phone call may relieve the necessity of visiting the person. The digital revolution also reduces the need for travel. Technological advances including the wide introduction of broadband mean that teleconferencing is becoming more common. This may reduce the demand for travel to and from conferences. Similarly, the use of virtual reality when seeing plans as well as the use of laptops and remote computers might reduce travel. The use of email and smart phones can affect demand, since often people can also send pictures, which could affect the need to travel within the company. Sometimes people can now more easily work from home. This was shown for example in 2014, when severe weather tossed the railway track at Dawlish in the air and disrupted other transport routes. Whilst some potential passengers would have used alternative modes some would have stayed at home.

The substitutes for transport are not, however, confined to business purposes, the newer types of education including the Open University, the Open College and London University International degrees all reduce the need to travel to and from educational buildings. Demand for entertainment and sport, which accounts for a significant proportion of off-peak travel has long been altered by television and increasingly video recordings. Sometimes football or rugby supporters will go to the ground of the away club (e.g. if they support a London club who are playing in the North of England they can watch the match on large-scale screens in the London club).

Quality of Service

Quality of service has a great influence upon demand. Some aspects of quality of service, such as comfort whilst important, are difficult to quantify and therefore to assess their influence. It is, however, easier to assess the influence of overall time of a journey sometimes called lapsed time. It is generally assumed for example, that people travelling to and from work would not wish to spend more than one hour each way upon this factor. This, however, is an over simplification and may vary considerably from one region to another. Commuters around the London area will often take a much longer time to get to and from work than the North East of England.

Time itself can be divided into three components that is accessibility time, waiting time and in vehicle time. If a traveller wishes to make a journey from London to Yorkshire, it may take 35 minutes to get to Kings Cross Station, and then they have to wait 15 minutes at the station. The journey may then take approximately two hours from London to York and has to walk 10 minutes at the other end to arrive at their destination. The total overall time is 3 hours of which 35 minutes is accessibility time, 15 minutes is waiting time, and 2 hours is in vehicle time.

Many transport economists have suggested that waiting time and accessibility time are more important factors than "in vehicle time". We argue that this, however, is an over-simplification, as it also depends partly upon the type of journey being made. A person travelling on holiday with a large amount of luggage may be more concerned with carrying the heavy luggage to and from a railway station or at an airport terminal than with the time on the vehicle itself. However, a commuter carrying just a few papers may be much more concerned in the morning or evening with the overall or lapsed time.

Why Valuation of Time Might Differ Between Modes?

The use to which the time can be put may also vary between modes and even within modes. Many commuters on the various rail services and business people travelling by air may be able to use this time for business purposes whereas this will not be possible by car except in the comparatively rare case of chauffeur driven transport. The rail companies have capitalised on this by providing seats where laptops can be plugged in so that in effect the train becomes an extension of the office. Free Wi-Fi is a great attraction to many passengers, especially younger ones. The amount of waiting time will depend partly upon the reliability of the service and partly upon the publicity given to times of departure. The greater the time between vehicles, the more important reliability becomes. Few people travelling in Central London in the rush hours will know the timetable for the Underground service. This is not important, since the assumption is that another Underground train will be there within a few minutes. If, however, one is waiting in unpleasant weather at a bus stop then the knowledge that the bus service is only hourly but not very reliable will increase waiting time very considerably. One of the problems of transport policy is that because there is often no overall effective control of transport policy that gains in vehicle time such as improvements in air services may be offset by changes in other modes of transport. The waiting time may be reduced in the future, as it is possible through radio control and closed-circuit television to indicate the timing of the next vehicle to potential passengers.

Safety

Whilst safety is an important factor in transport, perceptions of safety are perhaps even more important in determining travel mode choice. The publicity given to the occasional rail or air crash is far greater than the daily toll upon the roads. The publicity for example given to the Hatfield rail crash in October 2000 had an effect on the population. It also had an indirect affect since for a long period afterwards there were speed restrictions, which affected the quality of service. It was even odder with the Heck crash in February 2001 when a motorist went through a traffic barrier and landed on a line, which killed several people. The railway press commented that there were no restraints on the road traffic after this, but there were restraints on the railways.

Comfort

Comfort will tend to be more important the longer the journey time and will be an important factor particularly for the upper income groups. At weekends, the demand for comfort is usually less important and therefore currently the difference between first class and standard class fares on many routes is often small for travelling in a first-class carriage. Since it is often difficult to segregate passengers on different types of seats on the same train, some companies have abandoned the distinction. Recently, however, in the Kent area, the local company has re-introduced first class, but with a slight physical barrier between the two classes. The supplement is much smaller than the traditional 50% premium for slightly longer distance travel. This is presumably because the utilisation of such seats has not been very good.

Time Reductions and Effect on Demand

Improvements in the quality of service lead to faster overall journey times such as the electrification of the East Coast line between London and Edinburgh have led to increases in demand. The electrification from London to Edinburgh has led to a considerable increase in the number of commuters using the train service each day from Peterborough to London. The development of the motorway system has also led to a considerable increase in traffic particularly between Bristol and South Wales. The impact is not confined to domestic transport, as the demand for tourism has grown considerably with increasing speed of aircraft.

Status and Transport

Status affects demand for transport not only between modes, but also within modes. This is very noticeable with cars where a great deal of car advertising is geared to the status of the individual using the particular car. One bus operator in the West Midlands commented that some people would not admit that they used the bus to travel to and from work each day because it had such a low status. The National Bus Company before it was privatised carried out a series of advertisements particularly in the late 1970s in order to improve the status of its coach services and partially succeeded in this. In the USA, the Yellow Bus travel for school has the opposite effect and it is regarded as low status for a pupil to be driven to the school rather than coming in on the Yellow Bus.

Demographic Trends

The size and age of the population and the size of households will influence demand for transport. The age of the population will be particularly important in determining demand for educational travel as well as for business travel. Even for non-educational and work travel, the age of the population is important, children are unlikely to be able to carry out many journeys independently of their parents. Young married couples without children may well have a different pattern of social travel and shopping journeys from similar aged couples without children. A country such as the UK, which has many nuclear families (i.e. typically husband, wife or partner and a small number of children), will probably have a different pattern irrespective of income from countries, which have extended families. The extended family is where husband, wife, parents, parents-in-law, aunts and cousins will all live either in the same house or very close to each other.

Urban/Suburban/Rural Split

The demand for passenger transport depends partly upon total population. It is also affected by where they live and carry out their business activities. Countries, which have almost an entirely rural population, perhaps because they are dependent upon agriculture or forestry will have relatively little demand for passenger transport for journeys to and from work. In a typical modern developed Western country, however, such as the UK, the concept of rural dweller is more difficult. A comparatively small part of the total population, less than 2% is involved directly in agriculture though a much higher proportion of people live in rural areas. Particularly in rural areas near major cities, such as Birmingham, the people may live in a rural area but still depend heavily upon the nearest conurbation for their entertainment and thus demand for transport to and from the conurbation for their social journeys. They may also depend upon the conurbation for employment. In the more remote rural areas however, such as in Powys in Central Wales, car/van ownership is very high at 82.5%. This will not be because of demand for journeys to and from work, but because of the lack of public transport and the high cost of any remaining public transport. The suburban areas are more difficult to serve by public transport than more congested urban areas. The roads generally are less congested and therefore there is less of a deterrent to car ownership. Urban areas such as London, have a lower percentage car ownership than other areas with a comparable income.

Journeys to and from Work

For short distance journeys to and from work, the mode used will depend partly upon the availability of cars and partially upon fares and partly upon congestion and the ease of parking. For short journeys unless the rail service is very frequent the lower accessibility time of bus services will often outweigh the advantages of rail. For medium and longer journeys however rail services may be more important since the reduction in "in vehicle" times which can generally be obtained will outweigh any disadvantages of accessibility time. From Peterborough to London which is about 120 km (77 miles) the majority of commuters will travel by rail since road journey times whether by car, bus or coach will be appreciably greater.

In 1978–1979 journeys to and from work accounted for 18% of all journeys. Journeys by car accounted for just half of this with bus services accounting for 16% and rail transport 4%. Since that time, there have been considerable improvements in some of the rail services noticeably in the West Midlands with the Metro system. The majority of journeys to and from work were still of short length, about one third were under two miles. This pattern is not true of an area such as London. The proportion of people in Central London using public transport is far greater, for example in 1986 British Rail accounted for just over 40% of commuter traffic coming in the peak period into Central London. The demand for different types of transport may also depend partially upon employers' requirements. The increasing introduction of flexi-time or staggered hours could affect the timing of journeys to and from work as well as the modal split. The use of flexi-time, where people can work at hours which they choose usually around a central core time, enables passengers to choose transport services, which might be slightly less reliable but are otherwise advantageous.

Transport Technology and Effect on Demand

There has been a tendency towards faster trains particularly in France where the faster speed of the TGVs, means that there has been an increase in demand for longer distance rail journeys. The TGV service is currently being extended within France and to neighbouring countries. Similarly, Japan has had a considerable increase in its high-speed network.

In shipping, there has been a tendency towards catamarans on the short sea routes, which are faster than conventional ships. On aircraft, the tendency has been towards larger rather than faster aircraft so that journeys are likely to be cheaper. The European Airbus and Boeing's Dreamliner will compete for this traffic.

Journeys in the Course of Business

These are usually longer than journeys to and from work. Because the journeys are often paid for by companies rather than from individuals, demand, is likely to be relatively price inelastic and the quality of service, particularly overall journey time is likely to be one of the most important factors influencing demand. In the 1960s, the electrification of lines from London to Manchester and Liverpool increased British Rail's share of the market and adversely affected air transports' share. In the 1980s, the electrification from London to Edinburgh had the same effect. The West coast modernisation programme (from London to Glasgow including services to and from Liverpool and Manchester) now completed will it is hoped by operators including Virgin have the same effect of reducing air share of the market which is considerable. Unlike the USA, where there has been some demand for executive coaches there has been relatively little development of this concept in the UK. Budget airlines have become increasingly important. The Channel Tunnel has affected the demand for inter-city traffic between London, Brussels and Paris since the overall journey times by rail have become reasonably comparable with air travel between main centres. This has become even truer with more time taken for security at many airports. The Channel Tunnel acts as a substitute for air travel. Both formerly British Airport Authority (BAA) and formerly British Airways (BA) would have liked the Channel Tunnel link to extend to Heathrow Airport. This may sound odd but BA and BAA would have preferred scarce slots at Heathrow to be used for journeys for which there is no obvious substitute.

The airlines would be very interested in the idea of a high-speed link, perhaps going from the Channel Tunnel Link terminal near Kings Cross, St. Pancras to Scotland. This is because currently it is quicker for passengers to go to and from Brussels or Paris than it is to go from London to the North of England. If such a system were to come into being, then the overall time from the North of England to places such as Lille, Paris or Brussels could well be comparable to air, taking account of the likely waiting time at airports. An alternative (i.e. that the Eurostar services) could operate as both a domestic and an international route would seem to have advantages and Deutscher Bahn the German nationalised railway line would like this. However, currently the migration crisis would prevent this. In many other countries, it is often difficult to know which country one is in on a rail service since there are no obvious borders, particularly within the European Union, but this is altering because of terrorist attacks.

Demand for Educational Travel

This can be sub-divided into several categories. The demand for primary school travel is usually short distance travel in towns and cities mainly using walking or possibly accompanied car journeys. However, there has been some closures of primary schools in rural areas and therefore some form of public transport may be necessary. In recent years, far more children have been taken to primary schools by car (e.g. Beijing), this has posed problems of safety for other people as well as contributing to congestion. Sometimes the so-called "Walking bus" has been used so that children can walk in a group to and from school with adults supervising them.

The demand for secondary school public transport, however, has been much greater especially from the 1960s with the growth of larger schools and sometimes sixth form consortia. In March 2007, the then Labour government announced that it was extending free travel to and from secondary schools if pupils lived more than 2 miles from the school rather than the traditional 3 miles. This was part of an effort to cut down on the number of cars being used for such journeys. In the Further and Higher Education Sector, technical colleges and to a lesser extent the former Polytechnics and Colleges of Higher Education attracted a large number of people travelling on a daily basis. On the other hand, most universities especially those which have campus sites often have a smaller number of people travelling any greater distances on a daily basis though there may be a considerable degree of demand for passenger travel to and from the universities at the start and end of the terms. There is also now a demand at weekends to and from home for those students who do not live near the university to go back and forth every day.

Sometimes because of the absence of grants, students are perhaps more likely to go to a local university than in the past. The public transport operators were aware of this and former British Rail and National Express offered student cards which reduce the cost of such travel. In Birmingham part of the logic of the cross-city line was to provide direct access to Birmingham University so that students and lectures would be less likely to use cars. In another part of Birmingham, the University of Central England has a station directly outside one of its campuses. In general, however journeys for educational purposes are shorter than for all journeys. The journeys for educational purposes may in some countries form a very large proportion of the total, in Trinidad in the early 1970s it was estimated that about one third of all public transport demand was for educational purposes. In order to cater for this the schools had a shift system from 07:00 to 13:00 and 13:00 to 19:00. The evening peaks for education do not necessarily coincide with the evening peaks for commuting. In London, children under 11 are given free travel on the Tube, DLR and London Overground, as long as they are accompanied by adults (note: free travel alone only on bus and tram without an adult).

Demand for Shopping

Journeys to and from the shops have changed partly because the proportion of small shops has declined and this has been paralleled by an increasing growth of supermarkets and hypermarkets. The total demand therefore measured by passenger kilometres has grown substantially. There have also been more shopping centres, such as Lakeside in Essex and Bluewater in Kent, which attract people from long distances. Other factors, which have influenced the demand for passenger transport has been the growth in the number of refrigerators and freezers, which mean that shopping trips can take place less frequently than in former years. The increasing number of cars and in particular the increasing number of women who have passed their driving test has both altered the number of trips and also in turn influenced the location of shops. The effect of cars, however, should not be overstated. A research conducted by Dr Mayer Hillman has shown that even in the affluent South East of England where car ownership is high that a large number of people still use public transport or walking for journeys to and from shops.

It is not clear how far the use of internet shopping where people can buy groceries from larger supermarkets will affect the total demand for travel to and from shops.

The Demand for Hospital and Medical Services

From the 1960s onwards, general practitioners often formed group practices, which has meant slightly longer trips for patients and possibly their escorts. In the 1960s as well, there were often moves away from so-called cottage hospitals (i.e. small hospitals towards larger hospitals) which meant that people who were reliant on public transport had often much greater distances to travel. This was especially difficult on Sundays when public transport is often minimal. There has, however, been a partial reversal of the moves away from cottage hospitals since the 1980s. The demand has sometimes been met by special bus services, sometimes by diverting ordinary ones and sometimes undertaken in a voluntary capacity by special car drivers.

CHAPTER 3

Demand for Freight Transport

Several factors, which influence demand for freight are similar to those for passenger transport.

Changes in Land Use Patterns

The price of transport will influence the demand, but the price paid for the transit is only part of the total distribution costs, which is a more important factor influencing demand. Economists have often assumed that the demand for freight will grow in line with Gross National Product (GNP) and this is usually true for short periods. Changes in land use patterns particularly the growth of suburban areas and the moves towards light industry originally during the interwar years along the arterial roads from London such as the Great West Road affected demand. There was often ribbon development along such roads, so it was quicker to get to and from these factory sites, than to others, which might have been nearer, but were more inaccessible such as those in the more densely populated parts of town. Increasingly also modern light industry on industrial estates being found in areas around towns will influence both the total demand and modal split. Modern light industry will require less tonnage since the items themselves are lighter as well as the raw materials, than for heavy industries such as the steel industry. This will apply both to the journeys to and from the modern factories. Often the industrial estates are not linked to rail sidings, whereas usually heavy industry, including the steel industry was rail connected. Therefore, transport economists might expect to get a reduction in tonnage, but possibly more vehicle trips.

The Separation of Land Uses

The separation of housing from industry and the use of warehouses near the motorway intersections will alter both the total demand and the origin/destination matrix. The changing pattern of shops, particularly the growth of out of town superstores, will tend to reduce the number of trips, which need to be made by road haulage though this will not necessarily influence the number of tonne kilometres. The advent of relatively new regional shopping centres such as Lakeside in Essex and Bluewater in Kent will alter both passenger and freight demand. It remains to be seen how far the use of internet shopping including that for food will affect both passenger and freight demand. Will people who are not shopaholics wish to travel to and from shops, if they can get the same goods without the need to travel without the inconvenience of public transport or the problems of parking?

Historical Trends in Freight Demand

In subsistence economies, which is where people basically produce most of their few own sources of food and clothing, people will be able to produce the bulk of their own requirements or else will buy goods at local markets. Often transport costs are often high compared to total income and therefore the main demand for freight transport will come from the richer people within the community or generally for high value products such as spices where the costs of transport will form only a small proportion of total costs. This is known by economists as the importance of being unimportant. In most subsistence economies, fuel is likely to be found locally such as local timber or brushwood. However, as the price of oil has risen in real terms and oil is out of the reach of the poorest families in the Third World areas, women usually have to go further and further to get basic fuel such as brushwood.

Increasing use of coal in the eighteenth century meant that demand for both coal and iron ore, was one of the important elements of freight demand. Most transport demands were originally met either by water transport or railways since the cost of transporting the great weight of coal by horse would have been far greater.

Specialisation and Transport Demand

The increase in specialisation, means that currently there is a much higher percentage of both imports and exports. In Britain typically, about 40% of GNP is imported and exported and even goods which are produced in this country may well have to travel long distances. Environmentalists have used the term food miles to express their concern over this. This is, however, an oversimplification as often the amount of fuel to get crops to grow at unseasonable times may more than offset the amount of fuel used to transport the crops to the market.

Footloose Industry and Effects

In the twentieth century in Britain, most had become footloose, that is not tied to any particular place or location. There has been an increase in substitution of oil for coal. Coal is relatively expensive to transport and therefore most of the original heavy industry had grown up around the coalfields in order to minimise transport costs. There has been a decline in the British Coal Industry. This has meant that firms have been much freer to locate where they wish and if they are bulk-increasing industries as with parts of the furniture industry and brewing industry, they are located nearer the market. The substitution of oil for coal had significant effects for former British Rail, which originally obtained a large percentage of its total traffic from such carriage. In turn, this has also had an effect for the privatised rail industry. In the USA, however, demand for coal traffic is still a major part of total rail traffic, although this will change with many countries committed to reducing the effects of climate change.

Development of the Road Haulage Industry

The second factor was the development of the road haulage industry, which, meant that firms could more often be located on the outskirts of town where land was cheap. Originally, most manufacturers and many other companies had located either near the railways or along the banks of major rivers, such as the Thames or the canals. Nowadays, many depots are located near the intersections of main roads including the motorways.

Growth of Light Engineering Industries

The third factor has been in the growth of light engineering industries, which now includes the manufacture of computer parts. The transport of such items is relatively cheap compared with the value of the goods, but the finished goods are easily damaged and therefore firms may wish to move nearer to the market. Apart from the growth of light engineering firms, there has also been a tendency towards an increase in service industries. The demand for freight transport is generally less than that of the original heavy industries.

Different Measures of Demand and Different Trends

As with passenger transport, demand is measured in a number of different ways such as tonne kilometres, tonnage, vehicle kilometres or vehicle trips. These may show quite different trends. For example, during 1976 to 1986 the volume of tonne kilometres increased by 23%, whereas goods lifted fell by 2%. This is partially accounted for by the increasing length of haul for road haulage waterways and pipelines. About one third of all billion tonne kilometres in 1986 were the carriage of petroleum products, whereas transport of solid mineral fuels accounted for just over 6%.

Inland freight transport in Britain is predominantly road based which accounted for about 80% of total tonne kilometres, rail accounted for 8%, water by 7%, and via pipeline for 4%. However, this is not necessarily typical of a developed country. In the former Federal Republic of Germany road accounted for only 50% of billion tonne kilometres in 1985 with railways share being approximately 29% and water share approximately 19%. This partially reflects the importance of the River Rhine and partly the heavy subsidies, which were paid by the former West German government to subsidise rail services. In 2005 in the UK road freight accounted for 64% of all goods moved, rail for 9%, water 24%. Pipeline movements have remained static, at around 11 billion tonne kilometres. According to the latest data from the Department for Transport (DfT), road freight accounted for 76% of all goods moved in the UK in 2015, with the remainder by rail for 9%, and decreased by water for 15%. Total goods moved in Great Brattain accounted for 201 billion tonne kilometres in 2015.

The Forecasting of Freight Transport Demand

This may be very important for local authorities and the DfT, when considering the wear and tear on the roads. It has also been important for special enquiries such as the Armitage Report in 1981 on heavier lorries as well as the Wood Enquiry in London. To use a standard regression method for long term forecasting may be inappropriate, since much depends upon the assumptions used. One of the hypotheses was that as more trade occurred within the EU there was likely to be much greater demand for both imports and exports of manufactured goods between these countries and there will be less emphasis in producing bulk primary commodities. The British decision following the 2016 referendum to withdraw from the EU adds a degree of uncertainty about the future. However, part of the uncertainty also arises with a lack of knowledge of the total North Sea Oil resources. Oil accounts for about 60% of total demand.

Assessing Road Freight Demand

The proportion of vehicles with very heavy axle loads is of particular importance when assessing wear and tear since the usual assumption is that wear and tear on the roads increases in proportion to the axle load to the fourth power (i.e. that doubling of axle weight will increase wear and tear sixteen times).

The number of road haulage movements will depend partly upon the average size of lorries, which would have increased since the increase in permitted total weight to 38 tonnes in 1983. However, the maximum size is not the only consideration since the Armitage Report showed that relatively few lorries were limited by weight and far more were influenced by the size.

The number of lorry movements was also influenced by the load factor and also whether consolidation can be carried out. The load factor is the:

$$\frac{\text{Total load actually carried}}{\text{Capacity}} \times 100\%$$

Some firms notably Marks and Spencer have used modern management methods to drastically reduce the number of lorries serving their shops. There is no reason why other shops selling similar products should not do the same. Some commentators noticeably Terence Bendixson in his book 'Instead of Cars' (1974) have suggested that consolidation or having transhipment depots as in the Netherlands could considerably reduce the number of lorries within town centres. There is also considerable scope for reducing the amount of empty running with back loads. There is a particular problem for international road hauliers though this was overcome with a single internal market in 1992, the extension of the EU to 25 countries in 2004 and to 27 in 2007 and 28 in 2013 but 27 from 2019 (i.e. Brexit). The new countries are mostly from the former Eastern European countries from the communist bloc but also include Malta and Cyprus, which will have extended the freight market still more. The use of computers may help to reduce empty or below capacity running. Even at a lower level, magazines for the freight industry, which extensively advertise back loads will also produce an incentive to reduce the amount of empty back loading. Freight forwarders may be able to help to find return loads for lorries, especially where there is a regular demand.

Transport Technology and Its Influence on Demand

In the nineteenth century, the development of refrigerated ships meant Australia and New Zealand were able to export many products particularly New Zealand lamb, which would not have been possible before. The development of steam and iron ships with their lower costs compared with conventional sailing ships meant that exports of grain from the USA to Western Europe including the UK became possible. In the twentieth century, the development of roll-on roll-off ferries, containerisation, palletisation and packaged timber also improved the quality of service as well as reducing costs and thus stimulated demand particularly for international transport.

Although not a technical development as such, the use of the International Standards Organisation (ISO) size container, which is 8′ × 8′ by a multiple of 10′, led to a vast increase in containerisation. The containers could be transferred quickly from one mode to another and it indirectly led to the use of cellular container ships with a much lower handling cost. More recently there has been a move towards bigger containers, which has posed some problems for the UK rail industry with its limited loading gauge on many routes. There have been some improvements to this particularly for routes to and from Felixstowe, which will be able to accommodate the newer 9 ft. 6 inches containers.

The use of specialist conveyors for grain as well as quicker handling of bulk items whether bulk ore or oil has led to development of larger ships but have also increased handling speeds.

The digital revolution, will also by better scheduling, reducing the number of freight movements. This is especially true, if modern management techniques of operational research including queuing theory and network analysis can be used, which would help to not only minimise movements, but also to determine the best location of depots, the number of and type of vehicles which can be used.

The use of computers also helps both customers and operators to be kept up to date on transport demand as well as reducing the risks.

Total Distribution Costs

Total distribution costs will affect demand for freight. Total distribution cost will comprise the charges made for the transit, insurance costs, documentation costs if not paid for by the carrier, costs of packaging, risk of deterioration, damage or theft whilst in transit and the cost of holding stocks.

Air transport can compete for freight in spite of its generally higher transit costs, as it is often possible through its use to reduce total distribution costs especially the cost of holding stocks.

The cost of not having stock to satisfy the customer is difficult to determine. It depends partially upon whether the customers are willing to await a later delivery or alternatively whether they will look for another source of supply. If they look for another source of supply for one item, the danger may be that they will continue to buy this out of habit and that they may also buy further complementary goods or products.

The Cost of Holding Stocks

The cost of holding stocks may be expensive but is not a linear function of total stocks held, since the cost of a warehouse will not vary at least in the short run according to whether it is fully utilised or not. However, generally economists can assume that the cost will increase according to the amount of stocks because of the opportunity costs of capital foregone. There may be also greater problems of retrieving stocks the more that is held. This is analogous to the problems in an ordinary house of trying to obtain items which are lodged at the back of a household garage. The same problems can arise particularly in older warehouses. On the other hand, there may be important consequences, if stocks are not held for spare parts. The firm holding stock will not only have to take these considerations into account but also the possibility of obtaining discounts on large orders as well as changes in seasonal prices. Bulk discounts can be very large and if the firm has its own fleet of lorries, there may be advantages of large consignments which fill up the lorries rather than having half empty vehicles. There are the fashionable items (e.g. clothes, phones, including phone apps), there may be disadvantages of holding too much stock, since the items may no longer be sellable or only at a very reduced price. The advantage of a quicker form of transport such as air transport means that it is possible to minimise the risks of a stock-out without holding such vast quantities of stock. It does not necessarily follow, however, that air transport should always be used in these circumstances. Sometimes it may be sensible to use surface transport for most deliveries, but to use air transport for particularly urgent consignments. The same considerations may apply when choosing within a mode of transport, having regular deliveries even if they are fairly slow at cheap rates and paying a premium if necessary for more urgent deliveries. Industry has often moved to just in time production (JIT). This by definition helps to reduce stock but may be disadvantageous if for any reason supplies are disrupted as for example happened with the fuel protests in the UK in the early 2000s. This is in contrast to the other idea of just in case (JIC) (i.e. organisations stock items since they may be useful to customers). Habit is important. If people do not find a particular item in stock, they may turn to other supplier not just for that item but also for others. Producing firms have traditionally preferred JIC for non-perishable items, since it gives a more even flow of production and there will be economies of scale.

Containerisation has helped to minimise insurance costs. The insurance premiums have often reduced very dramatically compared with conventional break bulk handling. The use of containerisation and other unit loads may sometimes reduce the need for packaging. With air transport, it may be possible to send clothes with less packaging because of less risks of handling and documentation costs may be reduced with facsimile copying and emails means that there is the potential to notify the carrier consignee of details of consignment at the same time. The use of computers has been widely used to reduce the need for documentation. The SITPRO was the UK's trade facilitation body and was dedicated to "Making International Trade Easier" which was part of the Department for Business, Enterprise and Regulatory Reform that spent a considerable time in devising systems of aligned documentation. This made it much easier for firms to check their orders. International Organisations, such as the International Maritime Organisation (IMO), have tried to develop the systems requiring less documentation. In air transport, the use of the IATA waybill also has helped.

However, modern technology has the disadvantage of hacking where people can be sent false information about deliveries.

The Arbitrary Classification of Demand in Both Passenger and Freight Transport

When considering freight and passenger demand the distinction between passenger and freight transport may be somewhat arbitrary. People travelling in their cars to and from supermarkets will be classified as passenger transport even though the basic demand is for the carriage of freight, whereas shops delivering to customers is classified as freight even though in both cases the same function is being fulfilled. The use of pipelines for carrying oil is classified as part of total transport demand whereas the use of pipelines for the carriage of water or for gas is not regarded as part of transport.

If a parent takes their child 3 km to school by car, then the total passengers km shown will be 9 passenger km, since there will be six passenger km on the journey to school (i.e. parent and child were on the return journey to home), if the parent is travelling alone it will be another 3-passenger km. If the journey was done by bus or the person walking or cycling on their own, then the total passenger km shown would only be 3 passenger km.

Price Elasticity of Demand for Freight

How far demand for freight transport is price elastic or price inelastic depends partly upon the proportion of money spent on transport compared to the value of the goods. Generally, it can be assumed that the demand for transport of low value high bulk goods such as aggregates is much more likely to be price elastic than the demand for high value low bulk goods such as spirits.

Cross Price Elasticity of Demand

As with passenger transport, transport economists would expect to find differences between the short run cross price elasticity and long run cross price elasticity especially between different modes of transport. In 1979 following the Iranian Revolution, the cost of fuel grew rapidly and therefore road haulage costs rose compared with rail transport. There may be no immediate transfer to rail, partly because of the limitations of the railways to carry additional traffic and partly because in many cases people may have a contract with particular road haulage firms. The cross-price elasticity of demand is an important feature, if the government wished to become less reliant upon fossil fuels. Often firms' knowledge of total distribution costs has been fairly minimal according to surveys and therefore if firms are willing to look at alternatives the cost price elasticity of demand is probably very low. The late Peter Drucker (1909–2005) a well-known management expert, stated that distribution was the last frontier of management. Some transport economists believe that a sudden very large increase in price will probably have more effect upon demand than a gradual increase over a number of years.

Income Elasticity of Demand

The assumption has often been that demand for freight transport has an income elasticity of approximately one (i.e. that freight demand will grow in line with gross national product).

Lead Time

This is the time taken between placing an order and receiving the goods. The faster forms of transport will reduce the lead-time and therefore lead to lower stock levels being necessary.

The greater the number of orders which a firm places the larger the amount of administrative costs. The digital revolution helps minimise administrative costs. However, firms buying goods may well find that there are more substantial discounts the larger the order and therefore this will need to be taken into account.

Safety Stock

The minimum stock level needed to avoid problems of not being able to fulfil orders would depend upon the average level of demand. For some goods, demand may be reasonably well known and therefore the amount of safety stock is fairly easily predictable. If 100 units are demanded a week and this is reasonably certain and the lead time is four weeks, then a safety stock of 400 will need to be held. In practice, there may be slight variations and it is possible to use statistical methods to determine the chances of stock out with different levels of stock.

Obsolescence

This is less likely to occur with some industrial items than with consumer items. A firm supplying clothes or the latest mobile phones will face a very strong risk of these becoming obsolescent if it holds too high stocks. On the other hand, firms supplying ordinary electrical plugs will find that there is very little risk. Further factors determining the level of stocks will be that of the risk of deterioration or damage. Deterioration is less likely, if goods are held in warehouses where the environment is less likely to be damp or temperatures are more unlikely to alter.

The cost of holding stock will also depend upon the interest rates. Where interest rates are very high and inflation rates very high as in the late 1970s when inflation reached 25% the interest foregone was considerable. The reduction in the base rate in 2016 to 0.25% the lowest in recorded history in the UK aimed to reduce uncertainty following British withdrawal from the EU. In 2008, onwards with the credit crunch, it has been more difficult to obtain credit.

Sometimes the cost of holding stock is borne, not because of resale to customers but because the items are required for the firms' own production purposes. The cost of not keeping stock may, however, still be very relevant. Road haulage firms themselves in more remote areas may need to keep stocks of basic spare parts. The cost of not holding spare parts may be considerable. Increasingly where tenders are given there may be penalties from not being able to fulfil a contract within a certain time. Organisations need to consider, this when deciding their stock levels.

Why Air Transport May Be Used?

One of the reasons why air transport is frequently used in spite of its higher cost is because of the reduction in transit time. Less money is held with less stock. There is a shorter period before which the money will be repaid. Because generally air transport is faster, insurance premiums may be lower, particularly for theft since less time is spent in transit. However, increasingly with shipping using unit loads such as containerisation, have been reduced. Air transport may be justified, if it can result in sales, which would otherwise not take place. This has been true of newspaper distribution to the Mediterranean countries though in 1988. The Guardian announced that it would be able to print some of its newspapers in Southern France, which would presumably reduce the need for faster transport. Higher amounts of money may be paid, if the item is perishable such as some fruit and vegetables.

Government and Local Government Influence on Freight Demand

The government may influence demand for freight by the provision of taxation and subsidies directly to the transport industry as well as infrastructure pricing. They can also influence it through changing attitudes towards land use planning and regional planning.

Pryke and Dodgson in their book "The rail problem" published in 1975 stated that heavy goods vehicles did not pay their total social costs. This was true at the time and therefore would have meant that greater volume of traffic would have gone by heavy goods vehicles rather than by alternative modes than if prices had reflected the true cost to the community. This was partially altered by the changes in the vehicle excise duty system in 1983, when lorries were assessed for Vehicle Excise Duty on their total laden weight.

Subsidisation of Rail Freight

Many governments have subsidised their rail freight services considerably (e.g. China). In the UK, the government in 1974 under the Railways Act 1974 Section 8 gave 50% grants towards the cost of private sidings. The firms concerned had to ask for this, not the then British Railways. This posed some problems; sometimes private sidings might be justified to serve a number of companies but could not be justified by any one of them. There have been far less applications under this system than for comparable sidings in former West Germany or France. The advantage from the railways' viewpoint is that private sidings means that a door-to-door service is possible which otherwise would be a prerogative of road freight. The Railways Act 1974 also made provision for rail freight only lines to be charged with avoidable costs. Therefore, on lines where there are already passenger services running British Railways could charge cheaper prices. The EEC now EU has stated that there should be a system of marginal social cost pricing for all modes of transport. If such provisions were to be implemented, then this could well alter the modal split. In 2006, there was concern that there might be a reduction in the volume of freight using the Channel Tunnel since there was a dispute over the pricing system for the Channel Tunnel. It was announced in 2007 that the charge per freight train would be reduced from around £5,300 to £3,000.

The Government and the Provision of the Road System

The government through its provision of a comprehensive motorway system which is not paid for directly will increase the volume of traffic. There have been suggestions that this could be altered and there is a short stretch of the motorway system on the M6 where vehicles have the alternative of using a toll system. The M6 Toll (i.e. the Birmingham North Relief Road) is a 27 mile, three-lane motorway near Birmingham, which offers shorter journey times through the West Midlands. Tolls may also alter the distances travelled for any group of consignments since road haulage costs will depend partly upon time taken as well as distance. Therefore, there has been a tendency to locate warehouses near motorway inter-sections rather near the centres of towns or on waterfronts. This may be one of the reasons why distances and therefore total tonne kilometres have increased dramatically, whereas the tonnage has not increased significantly in past years.

British entry into what was then known as the Common Market in 1973 and the creation of a Single Internal Market in 1992 means that there has been a considerable increase in traffic between Britain and the Continent. The increase in size of the EU to 28 countries in 2013 will have accentuated this development. This, however, has been combined with other developments such as roll-on/roll-off, so it is difficult to tell exactly which are the individual causes for the increases. It is not clear what the effects of British withdrawal from the European Union will be on freight transport.

The Demand for Fuel

The demand for fuel will both alter the total demand and possibly the modes of transport used, and is significant in the UK, which has a high demand for fuel. Currently more port facilities are available for liquefied petroleum gas.

The huge tankers, which transport this from Qatar, are a significant part of the shipping market. They will deliver this to Milford Haven and then often by pipeline to other parts of the UK. The advantage of transporting petroleum in a liquefied form is that it takes up substantially less volume at very low temperatures around −170°C and then the product comes out from the insulation so that it becomes a normal gas. In addition, it should be noted that Norway was the main source of both crude oil and natural gas for the UK in 2016. The Langeled pipeline built in 2006 and 1,166 kilometres long under the seabed transports natural gas from Nyhamna, Norway to Easington Gas Terminal in the East Riding of Yorkshire. More specifically, the UK imported approximately 43 million metric tons of crude oil from Norway (accounting for 61% UK total fuel consumption).

There has been controversy in the UK and in other countries about the use of fracking. This means injecting liquids at high pressure to extract oil or gas. Many environmentalists suggest that this is inherently dangerous. Others such as Lord Deben (a former Conservative cabinet minister) state that renewable resources have constantly reduced prices in real terms and therefore it is unnecessary to allow fracking.

Freight Transport Pollution

Although the government has announced new diesel and petrol vehicles and vans will be banned in the UK from 2040 in a bid to tackle air pollution, there is still a long way to go. Pollution has often been associated with heavy lorries mainly because of the problems of heavy black fumes emanating from badly maintained vehicles. The provisions of the licensing system since 1988 with its emphasis on the operator's record may help to reduce this. Vehicle pollution still kills many people. The use of lead in petrol leading to the problems of ethyl and tetra methyl lead has been widely publicised. The use of lead in petrol has been gradually phased out and banned in the UK from 2000. It has been difficult for car drivers to do much about this as car manufacturers and petrol companies between them had not been making lead free petrol widely available. However, changes in the road taxation system in 1988 helped to reduce lead in petrol. Another possibility would be to use diesel rather than petrol. Until recently, this was thought to be less harmful than petrol. Current research shows that this is untrue. Diesel is only generally cheaper for motorists if they drive large distances during the course of the year. One way of reducing both pollution and other social costs would be to reduce the number of vehicle kilometres through computer programming. This becomes easier for both road freight and road passenger as computers and associated software have become cheaper and have greater memory capacity. New York has moved towards having its taxi fleet being composed of hybrid vehicles and this is a model, which local authorities could follow in other countries. In addition to this, a growing number of electric-powered taxis have been rolled out in outer Beijing since 2011. Some insightful lessons can be learnt especially for developing countries.

CHAPTER 4
Cost Structures

This chapter deals briefly with the cost structure of the different modes of transport. The different definitions of costs are defined.

Fixed Costs

These are costs, which do not alter with the volume of traffic whether passenger or freight. Some such as the Channel Tunnel have almost all their costs as fixed costs, whilst others will have a much smaller percentage of their costs as fixed costs, and it is important for transport operators to realise the significance of this.

Average Fixed Costs and Average Costs

These can be defined as the total costs divided by the total number of units produced. These can be found by dividing the fixed costs by the number of the relevant units such as passengers carried or passenger kilometres. The average fixed cost curve is always the same shape, since the fixed costs will be spread over a greater volume of traffic. For the Channel Tunnel, the fixed costs are very large the average fixed cost (AFC) is very important, since only with a large volume of traffic will there be any profitability. Whilst originally the demand was forecast to be around 20 million passengers per year for many years the numbers were around 7 million which is one of the reasons why Eurotunnel has suffered many financial embarrassments. They have, however, now reached about 20 million a year.

Variable Costs

Variable costs alter with the volume of traffic. With almost all modes of transport, these will include fuel costs and the changing pattern of oil prices and alternatives will have a significant effect on many modes of transport. Variable costs of a transport firm can include wages payments such as overtime and possibly some bonus payments, which are based on productivity.

Average Variable Costs

This can be found by dividing the total variable costs by the number of the relevant units such as passengers carried or passenger kilometres. Similar figures can be found for freight.

Total Costs

Total costs are found by adding the fixed costs to the variable costs. The total cost line is parallel to the variable cost line. If a taxi has £3,000 fixed costs, then the total cost curve will be £3,000 above this.

Marginal Costs

Marginal costs are the additional cost of one more unit of output (e.g. another vehicle kilometre or another passenger carried). Transport economists can subdivide marginal costs into two categories. These are short run marginal costs and long run marginal costs.

Short Run Marginal Costs

These are the costs per unit of carrying and additional unit of traffic whether passenger or freight when capacity cannot be altered. The importance of this concept can hardly be overestimated in public transport or for car and lorry drivers.

Long Run Marginal Costs

These are the costs per unit of carrying additional units of traffic whether passenger or freight when capacity can be altered.

Suppose a school taxi regularly has to pick up five passengers to and from a school but it only has capacity for four passengers. Clearly, it will therefore have to make a double trip, since it cannot carry all five passengers at one time. So, in the short run, the costs of carrying the fifth person is quite expensive. This will include the driver's wages for the second trip as well as more fuel.

In the longer term, the taxi driver or owner could consider the additional cost of having the five-seater taxi rather than four. This is analogous to some airlines. They guaranteed that they would take passengers for a fixed sum and the aircraft had a capacity of about 100 people. If therefore 101 passengers turned up for the flight, then the 101st passenger would have been very expensive to carry. Therefore, the airline would willingly give the passenger a considerable sum in order to avoid the costs of having another aircraft almost running empty. In the longer run, the solution might have been to allocate a slightly larger aircraft to that route or perhaps to raise the fares slightly in order to avoid this embarrassment.

The railways sometimes running through bottleneck points have the same dilemma, although one solution might be to have standing passengers, since the capacity of trains is not clearly determined.

Passenger overcrowding has been a major problem and in March 2007 the government announced that it would allow for another 1,300 rail carriages, but it was not clear how soon these would enter into service. Currently, there is still overcrowding on some routes especially those from Reading to London.

Importance of Short Run Marginal Costs

In most of the transport industry, the unit of supply (e.g. the bus, coach, taxi or aircraft) is not the same as the unit of demand. Therefore, the marginal cost of additional demand will usually be negligible, if the vehicle is operating below capacity. An off-peak train, could carry more people at a negligible additional cost apart from a very marginal increase in the amount of fuel. Therefore, the railways will gain from any additional passengers, if they do not abstract from the existing passenger revenue. Because it is easy with aircraft to know when and where people get on and off, the idea of maximum yield from additional passengers has become even more important. The advent of the internet makes it easier to inform potential passengers whether any seats are left. This is not quite so easy with coach passengers, but some coach operators have used standby fares.

Importance of Long Run Marginal Costs

Transport economists looking at bottlenecks on the railways system, such as the Worcester to Paddington line, where the track was singled during the 1980s or just outside London Bridge Station where the lines for Cannon Street and Charing Cross diverge, will assess the total costs of removing the bottlenecks. The problems could be overcome, but it would be expensive to do so. This will be done with the Thameslink development where a very frequent link between North and South London is being developed and currently, London Bridge is being altered very considerably. Bottlenecks also occur on shipping routes.

Depreciation

Depreciation is not straightforward. It is the loss in value by the assets of a business because of their use during the course of the year. It is often called 'fair wear and tear'. For example, if we buy a new van on 1st January for £14,000 and use it for a year, it will not be worth £14,000 at the end of the year. Everyone knows how quickly motor vehicles depreciate and by the end of the year, it is quite likely that if we traded the van in and purchased a new one, we would not get much more than £10,500 as the trade-in value, possibly even less. However, it will also depend upon the mileage, since if it had lain mainly in the garage, it might be worth £10,500, whereas if it had been very much used particularly in town with constant stopping and starting which puts more work on the engine, then it might be worth only £9,500.

Economists could say that £3,500 was a straightforward fixed cost where the additional £1,000 loss in value was part of the variable cost. In practice, depreciation is often taken as a fixed cost.

Joint Costs

Joint costs occur where the provision of one service means that the cost for another service has also been incurred. The simplest illustration of this is return journeys.

The railway frequently has joint costs which means that a stretch of track is often used by both passenger and freight trains and by different types of trains including stopping and express passenger trains as well as freight trains. Since privatisation, it is slightly more complex, since the passenger trains are often run by different operators.

There is no straightforward method of allocating joint costs. There have been disputes about how to allocate these both under the nationalised and the privatised regimes. One way of doing it is simply to allocate costs according to the number of trains, but this would bear more heavily on smaller trains proportionally, although it could be argued that if the capacity is used up then this is reasonably fair. Sometimes it could be more sophisticated taking into account the wear and tear on the track. This may seem abstruse, but it is one of many problems besetting the Channel Tunnel at the time of writing since the costs for through freight traffic was increased. The Rail Freight Group at one stage suggested that if action was not taken there might not be any through traffic at all. This would be wasteful, since one of the major advantages of the Channel Tunnel is that it could take much of the long distance heavy freight traffic off the roads.

Allocating Costs for Return Journeys

If any vehicle goes out, there is not a straightforward way in which to allocate the outward and return journey costs. If half the costs were allocated to the outward and half to the return journey, then in the morning peak the operator would be charging the inward journeys much less per passenger and much less than on the return to the suburbs. The operator would probably find that they would be obtaining few passengers on the return journey when the additional cost of providing for them would be negligible. On the railways, the Cooper Brothers formula was used for allocating joint costs and it did suffer from this imperfection.

Perceived Costs and Objective Data

This is a very important concept in transport economics. These are the costs which people, particularly motorists, think it costs and on which they act. There is plenty of evidence both anecdotal and academic to show that people understate their true costs of motoring. Objective data can be obtained, for example from the RAC figures from 2016 which, reveal that the cost of running a vehicle is growing steadily.

Opportunity Costs

This is one of the most important concepts in economics. It means what do operators give up. Economists sometimes use the phrase "the next best alternative forgone". Often land values change considerably in the UK. Currently one of the limitations of conventional accounting is that it shows what has been paid but not what the market value for the land could be in another use. There have been examples of asset stripping, where firms have taken over another firm and then sold on some of the particular parts of the company for much higher values since they understood this concept. This particularly applies to land, which may have been used mainly for a parking lot by the company in the town centre or even in a suburban area. As the value of land has often increased in both money and real terms (i.e. allowing for the effects of inflation, firms will consider alternative uses for the land). A road haulage firm using the land for parking space might often mean that the firm has been able to make more profits by moving to other parts of town. Currently, there have been many comments about private equity firms taking over some of the largest firms in the country. Whilst this has not so far applied to the transport industry, it seemed likely that it could do so in the future, although the credit crunch makes this slightly more unlikely now.

Wages in the Transport Industry

Wages are mainly determined in the transport industry, as they would be in other industries. There are, however, a number of different emphases, which are more strongly felt in the transport industry than in many others. The first is that shift work is very common; for example, in ports ships can arrive at any time of the day. It will be easier for ports to cater for the liner traffic than tramp traffic (ships that carry low value commodities such as coal). Liner means routes which are scheduled for particular times and days. Therefore, assuming that there is a regular pattern, the ports will know well in advance when the ships are likely to arrive. Therefore, they can plan accordingly assuming that weather conditions do not prevent this occurring. On the other hand, catering for tramp shipping is more difficult. This is both because it is more difficult to know at what times the tramp ships will arrive, but also because there will be a strong cyclical pattern with far more tramp ships trying to arrive in boom periods rather than in a recession.

Ports and Casual Labour

Until the 1960s, in the UK, casual labour was often employed on a daily basis which was unsatisfactory for a number of different reasons. Part of this was that it led to favouritism so that if a family were in favour with the people selecting the dockers, then they would be much more likely to be chosen than otherwise. There was also the point that it discouraged any training since why should people bother to train when the likelihood of them being able to use the training was problematic even if when they turned up they did get a job. It also did not encourage people to be loyal to the industry, since why should people be loyal to the industry, if the industry seemed not to be loyal to them. The conditions in UK ports often compared unfavourably with that of other developed countries, such as the Netherlands where a great deal more work was under cover. The system was eventually abolished and the National Docks Labour Board set up. For the shipping industry itself, the distinction between tramps and liners is important. It is far easier to have regular employment with the liners since the demand for labour is known in advance and therefore ship owners can plan accordingly. With tramp shipping this is much more difficult.

Timing of Loading and Unloading of Ships

Whilst it could be argued that port owners do not necessarily have to unload the ships at any particular period, the ship owners do not wish to have their capital assets lying idle for a great deal of time. Until the 1960s, even on the longest distance runs, ships often spent more time in port than at sea. With more specialised ships, often costing proportionately more, there has been much more emphasis on trying to ensure that the ships spend time at sea where they are earning revenue and not in ports where they are not. This in turn means that the shipping and port industries have become more inter-related and with much greater mechanisation taking place the port owners themselves have been more interested in trying to get better value from their own capital assets.

Problem of Long Periods Away from Home for Some Shipping Personnel

Apart from the distinction between tramp and liner traffic on international voyages ships personnel will have been away for long periods at a time and this means that payments have to be made accordingly. The tradition of men getting large sums of money at the end of their voyage, spending it on drink is not acceptable to modern opinion and therefore ship owners have to try to devise other methods of more regular payments, if possible.

Problems of International Flights

Within the aviation industry which is almost by definition mainly international rather than domestic for the UK, although not necessarily within large geographical countries such as Australia or the USA, there are problems of trying to have sufficient cabin crew while conforming to the international flights regulations. For example, in the UK, pilots are not usually meant to fly for more than 900 hours per year. The problems of airlines are not perhaps as acute as shipping where there can be extremely long voyages, but there are similarities. When looking at wages and conditions of work, economists need to be aware of the problems that being away from home can cause. This is especially as more women enter the workforce and also some men are becoming concerned about lack of contacts with their families. One of the other features of airline recruitment is that pilots often retire early.

The requirements for airports are that at most international airports there are peaks, but there must be staff at all times, especially for security purposes. The peak hours differ slightly from those of other modes of transport. The basic pattern of shift work however is still common.

Coach Hours Regulations

Coach operators have to obey hours' regulations. The former 3820/85 regulations, have been replaced by 561/2006 regulations. The coach operator has to be aware of these, although they do not apply when the single route length is 50 km or below and then British domestic rules apply. The same rules apply to 9–16 passenger seats, if the buses are being used commercially and to all journeys with vehicles with 17 or more passenger seats.

There is a maximum driving limit of 56 hours as long as the total working time does not exceed the road transport working time regulations of 60 hours. There is also a maximum fortnightly driving of 90 hours in any two consecutive weeks. Whilst Britain is now leaving the European Union, it is unlikely that the hours' regulations will be altered.

Whilst driving hour regulations have always been imposed, there have been methods of fiddling the records. Digital tachographs have to be fitted to all new vehicles, which entered service from 1st May 2006. The rules on scheduling means there is less scope for altering hours, than would be the case in typical factories. With domestic bus services then there are problems of trying to ensure that the daily peak hours are covered in most urban areas. A split shift system is not usually acceptable.

Railway Drivers Hours

On the then UK railways there are the standard problems for drivers of passenger trains, although it is possible unlike the road sector for drivers to drive either freight or passenger trains. The railways in the UK also have additional requirements (i.e. that they require the drivers to have route knowledge). This means that there are problems when there are engineering works in being able to allow drivers onto possible divisional routes, which restricts the rail operators more than others. Pryke and Dodgson in their book 'The Rail Problem: An Alternative Strategy' (1975) reported on the various restrictions, which in practice were imposed so that often drivers would not be able to go through to one area to another. In the days of steam trains, there were often ingenious methods of getting from one part of the train to another, since usually train drivers in the UK unlike some other modes of transport and in some other countries normally go back to their home depot rather than staying overnight.

Privatisation has not helped since often drivers cannot usually drive on other operators' routes even when with diversions this might be helpful.

Problems with Bonus Schemes

Transport is not homogenous. This means that it is often difficult to devise bonus schemes, which reward people who work harder. There have, however, been a number of different bonus schemes in different modes of transport.

Punctuality Bonuses

Since one of the basic requirements is punctuality, bus operators have sometimes tried to reward punctual drivers. Clearly, since the road system is not usually within the direct control of the road operators this may cause problems. If there is a punctuality bonus system, bus drivers may be tempted to try to speed up loading at bus stops, either by not collecting fares, or by simply encouraging passengers to hurry up even if this increases safety risks. There is also the risk that drivers may speed and take risks. The same also partially apply to road freight, although there is usually more flexibility about routes.

Safety Bonuses

Safety is one of the main requirements of the transport system. If safety bonuses are rewarded, then there may be problems since minor accidents might not be recorded and this could lead to problems both with maintenance and to problems of lack of safety of vehicles later. Part of the problem is that whilst railway safety is largely with the exception of level crossings under the operators' or the track providers' control, this is not true of the road system. However, there can be an analysis of road accidents or any other accidents, which would indicate using statistical methods, such as the Poisson distribution, whether or not accidents occur more frequently with some buses or some drivers, than with others.

Proportion of Fares Bonus

Sometimes a proportion of fares have been allocated as a bonus. This has caused problems where conductor guards have been awarded these and they may not therefore be willing to accept valid excuses for not paying fares. There are particular problems where ticket machines are out of action, at un-manned stations or even where with manned stations the queues are so long that frustrated passengers will wish to buy their tickets on the trains whatever the operators may think.

Fare evasion is a constant feature of the railways, in particular therefore there could be consideration paid to a group, rather than to an individual bonus system. The other problem with proportions of fares being given is that in road passenger transport that drivers might sometimes ignore older passengers if they are not paying or be less willing to carry season ticket holders rather than the more irregular passengers who will pay fares. As more fares are paid off the bus, this is not a very good system.

Sometimes bonuses for freight traffic have been awarded according to the number of items handled. Before the container revolution, this was common. Sometimes workers were almost entirely paid on this basis, although some articles were much easier to handle than others, again reflecting the lack of the common measure of transport output.

A further method can be paying bonuses according to previous targets set. There was condemnation in the rail industry when Network Rail bonuses to the chiefs of the organisation were awarded immediately after the crash near Kendal in Cumbria, which killed an elderly passenger in 2007. Many commentators might feel that if the employers are going to award themselves a bonus, if things go right, perhaps they should be more willing to take direct responsibility, if there are fatal crashes for which the system as a whole needs to take responsibility.

CHAPTER 5

Forms of Competition

Economists give a variety of names to the different forms of competition depending upon how much competition there is in the market for transport services. These range from perfect competition at one end to monopoly at the other end.

Assumptions of Perfect Competition

There are a number of conditions required for perfect competition. These include a large number of independent buyers and sellers, perfect knowledge on the part of buyers and sellers, freedom of entry and exit for sellers, no undue discrimination on the part of buyers and sellers and factor mobility. The phrase no undue discrimination means that consumers will pay the cheapest price, which is compatible with the quality of service that they require. If buyers and sellers have perfect knowledge, then there cannot be different prices in one part of the market to another since if the prices are higher, then the consumers will not buy them, and equally if the prices are lower, the sellers will realise this and will raise their price.

The term independent buyers and sellers are required since if many of the sellers of service get together they could form a cartel. Thus, it was often claimed that shipping conferences formed this. Shipping conferences no longer exist. It may also be important on the buying side since Oil Petroleum Export Countries (OPEC) countries formed an effective cartel for selling oil. This therefore means that the oil market is by no stretch of the imagination perfect.

The unexpected very sharp fall in oil prices in 2015 has shaken this assumption that even an effective cartel can always maintain prices of the goods or services that it sells.

In the short run, the firm can make supernormal profits, which can be defined as those which are above average for the industry, normal profits, or losses providing in the latter case that it is more than covering variable costs. Normal profits are defined as those, which give no incentives for firms to leave the industry or others to join the industry. Under certain stringent conditions, perfect competition will lead to a Pareto optimum. The Pareto optimum means that no consumer could be better off without making another consumer worse off.

In the long run, the assumption is that firms, which are making losses, will get out of the industry. There is no possibility of long run supernormal profits, since new firms will enter the industry. Supernormal profits are defined as those which are above those, which are expected in the industry.

Tramp Shipping as an Example of Perfect Competition

As with other branches of industry, it is difficult to find many examples of perfect competition, though tramp shipping comes very near to being an example. The Baltic Exchange acts as a medium whereby information about rates is freely available and therefore there is reasonably perfect knowledge of the costs of tramp shipping. There are large numbers of sellers since tramp shipping is on a worldwide basis. There are equally large numbers of potential buyers of cargo space. Whilst there is sometimes discrimination on the part of certain buyers or sellers in the shipping market, for example embargoes on particular shipping lines, overall there is no great discrimination on the part of buyers of sellers.

Problems of Differences Between Perceived and Actual Costs

Perfect competition does not usually exist in transport, partly because perceived and actual costs can differ very considerably. This is particularly true of motoring where there is a considerable amount of evidence that people underestimate their costs. It is also true of own account transport, where there is evidence from Dr Clifford Sharp and others that firms do not know their total distribution costs.

Transport could be made more competitive, if people were aware of their true costs and this could sometimes be altered by government action. Raising the price of fuel would bring perceived and actual costs more into line. Sometimes it can be done through publicity, or better information can be provided through agencies such as the Baltic Exchange and by using brokers.

The various transport modes could be more competitive, if operators brought their perceived prices more into line with the marginal costs of competing modes. This can be done with the use of Travel Cards, Family Railcards, Disabled Card or group fares. Usually there is a certain fixed sum per period, usually a year, plus additional charges in some cases, which are similar to customer's marginal perceived costs.

Examples of Barriers to Freedom of Entry

Perfect competition is also infrequent in the transport industry partly because the conditions of freedom for entry do not always exist. Until the changes in the 1980 and 1985 Transport Acts in the UK, it was difficult for new firms to enter the road passenger industry. Until the 1968 Transport Act, it was difficult for new firms to enter the road haulage industry. In both road freight and road passenger transport, there are some restrictions on entry into the national market. In the past, the EU did not allow cabotage in either the road freight or road passenger market. Cabotage is where operators have not been allowed to both pick up passengers or freight in the same foreign country. An operator could have picked up goods from the UK to France and similarly could have picked up goods from France to the UK but could not have picked up goods from France to deliver to another part of France. This accentuated therefore the normal problem of trying to avoid empty loading on one haul of the journey.

The European Union altered the rules about cabotage in 2016 so that the minimum wage has to be paid on all journeys.

In some towns, it has been relatively easy to enter the taxi market, though in other cases, local authorities have imposed restrictions upon the number of firms. In New York, there have often been restrictions on the numbers of taxis operating which means that a licence for this purpose can be sold at a premium. This means that it is difficult to get passengers for both outward and return journeys, if they go outside the city boundaries.

In the aviation industry, there have often been limitations imposed upon new routes or new operators by the Civil Aviation Authority in this country and by international agreements for international air transport. In shipping, however, within the UK and many other countries there have been no restrictions on entry or exit. The rail market is difficult to enter partly because of land use but also because governments have usually imposed restrictions. Historically, the need to acquire land before the advent of compulsory purchase meant that it was difficult for many operators to find a route, which was not already occupied by a rival operator. This would have been more difficult in a country where there are hills, which meant that there are only a limited number of potential routes especially in the nineteenth century when most locomotives were not powerful enough to tackle any steep gradient. Usually a gradient of about 1–70 was regarded as severe. There is still a problem of entry, even though there is now a franchising system as the operator Grand Central Railway (now owned by Arriva) from Sunderland to London has discovered. It finally was allowed to run services after a judicial review of its proposals in 2006. The Great North Eastern Railway (GNER) Company, which then held the franchise from Kings Cross to Scotland, objected to the Grand Central Railway since it felt that it would lead to some passengers going from their services to the rival. They also claimed that this negated the logic of the franchise, which is that an organisation buying the franchise knew what the competition was.

There are often only a limited number of slots at some of the major airports, which means that new operators have often looked at other airports for their flights.

In the UK, the railway franchise system imposes great barriers to many potential railway companies because of the high costs of tendering. The House of Commons select transport committee in February 2017 stated that the present United Kingdom railways franchise system is unfit for purpose.

Lack of Perfect Information

The lack of perfect information is more likely to exist on the passenger side where there are larger numbers of customers than with road freight where there are fewer customers and the buying of freight services may well be undertaken by specialist people. On the other hand, passengers frequently do not have very much information and the large number of fares, which are charged on the railways with their numerous variations, makes it difficult for many people to be able to work out what the cheapest fares are. The railways in the UK did simplify their fares structure in 2008. There are, however, still currently criticisms that rail operators do not always give people the cheapest fares even when the passengers specify correctly what they need.

With buses, it is often difficult to find out what services run from one area to another. Often the operators publish maps, which only show their services, which are not very helpful especially as sometimes other operators would have bus routes, which could act as a feeder service to the main company. Sometimes by not listing other companies' services, it may mean that people do not take the journeys, since it would be helpful for passengers to know that there are return journeys at a convenient time.

Lack of perfect information about the market will mean that firms will not always enter the market even if it is profitable, and sometimes will not leave it even if profits in the economist's sense of the word are non-existent. This is particularly true where firms do not allow enough for replacement costs. This is very likely in periods of high inflation.

The lack of perfect information sometimes arises, since there are sometimes problems with physical co-ordination between and within different modes of transport. This can be seen in the usual UK problem of very few purpose built interchanges. This often makes it more difficult for potential passengers to find out where services are going to and from and in the case of disabled people or those carrying a great deal of luggage may mean that it is very difficult to make a journey without assistance. Bradford in Yorkshire has a purpose-built interchange. Other countries such as the Netherlands have not usually suffered from this problem and with purpose built interchanges. It is easy to go from one mode of public transport to another. The Amsterdam Central Station is a good illustration of this.

If transport organisations do not allow for replacement costs then unless there is an immediate liquidity problem they will not necessarily go out of business straightaway but they will tend to depress the total market level and therefore could jeopardise other firms because of short-term considerations. Economists realise that perfect competition does not necessarily lead to the optimum allocation of resources, if there are untapped economies of scale. It seems unlikely that this situation often arises.

Competition and Safety

Transport operators have often claimed excessive competition could lead to lack of safety. This has been claimed in a number of different modes, including road passenger, road freight and air transport. It was alleged in the 1920's that the so-called 'pirate buses' did not take sufficient account of safety. Pirate buses were those which challenged the established operators often by undercutting the fare charged. Sometimes there are anecdotes that they would cut out some intermediate stops so that they could run slightly ahead of their rivals. Similarly, the Salter Committee, which reported on road haulage in 1933, was told by the larger companies that safety was a problem caused by smaller firms' price-cutting. There seems to be little empirical evidence to support this statement. In air transport, it has been stated that the use of IATA standard fares was necessary to provide adequate fares and charges so that safety was maintained. There seems to be some evidence that the safety record on scheduled flights is better than that on charter flights. There may be, however, other reasons to explain this. There is little evidence that the abolition of the scheduled fares agreement on prices has had any adverse effect on safety.

There has been little evidence either in road freight following the 'O' licensing system introduction in 1968 that this has led to a reduction in safety. The Foster Committee in 1978 and the Armitage Committee in 1980 broadly endorsed the 'O' licensing system. There has also been little evidence that the 1980 and 1985 Transport Acts have led to any great differences in road safety. In the rail industry, there has been more concern about the role of subcontractors, which means that too much emphasis seems to have been placed upon detailed contracts and not upon the supervision of maintenance. In 2007, there were still problems about how far maintenance was adequate on the rail network after the crash in Grayrigg. Christian Wolmar has commented about the problems of safety in his book "Broken Rails: How Privatisation Wrecked Britain's Railways" (2001).

Oligopoly

A definition of oligopoly is that a few firms compete in a market. There are problems in determining output and price in oligopoly.

In all forms of competition, whether monopoly, oligopoly or perfect competition, profit maximisation will take place when marginal revenue equals marginal cost. The main problem for the oligopolist is trying to determine marginal revenue because of the problem of interdependence. Organisations will try to predict their rivals' reactions. The fact that the oligopolist raises or lowers his or her prices is likely to lead to competitors' reactions, which therefore in turn alters the operator's marginal revenue. It is often difficult for firms to have perfect information about how their rivals will react to their price increases or decreases. They are even less likely to know exactly how other firms will react to non-price competition aspects such as changes in quality of service.

This is not true of perfect competition or monopoly. In monopoly, there is by definition no need to take account of the reactions of rivals, since there is not one and in perfect competition the market price is given for the individual firms. In oligopoly, one firm altering its price may lead to a competitor reacting through price or quality alterations or both.

Oligopoly Models

A number of economists have attempted to look at the problems of oligopoly, one of the earliest was Cournot 1801–1877. His model was one where one firm altered its prices and was continually being surprised when one of its rivals reacted.

Game Theory

Perhaps the most sophisticated theory in oligopoly is the game theory. Competition does not necessarily have to take the form of price competition. It could take the form of non-price competition (e.g. changing the quality of service). Often competition between modes may take this form. Air transport will usually be more expensive than other modes and therefore the form of competition will have to be non-price.

Perfect and Imperfect Oligopoly

Oligopoly can be divided into perfect oligopoly and imperfect oligopoly. Perfect oligopoly occurs where the product or services is homogenous in the eyes of the consumer. Imperfect oligopoly occurs where the product or service is different. Perhaps the more basic distinction in oligopoly theory is where firms seek to compete and those where they seek to collude. How far collusion or competition takes place depends partly on the central government's attitude or other appropriate authorities, including the EU. There has been criticism in the UK bus industry that the government has often not allowed agreements for interavailability of tickets, which would have been in passengers' interest.

Within these broad headings, economists can further sub-divide into non-price and price competition. They could regard the old IATA arrangements as part of a collusive model for scheduled flights, though non-price competition such as advertising still took place.

In the bus industry, there have been Transport Acts, which means that details of joint services had to be informed to the Office of Fair Trading (OFT). This office was dissolved in 2014 and its responsibilities passed to a number of different departments. Collusion is generally more likely in a declining market rather than an expanding one. This is because transport organisations in an expanding one may well feel that reducing prices would help to give them an increasing share of an expanding market. On the other hand, with a declining market transport firms are likely to feel that the reduction in fares or charges can only lead to an increase in demand at the expense of other firms (the so-called zero sum game).

Advantages and Disadvantages to Firms If They Break a Collusive Agreement

A paradox of collusion is that the more that transport firms keep to collusion then the greater the rewards for any transport organisation, which can undercut the agreed price. If there were approximately 100 airlines, which kept to the old IATA fares structure, then any one airline, which could give secret high discounts to travel agents or supply more tickets to bucket shops, would gain a considerably increased market.

Bucket shops is the rather odd name given to agents who sold below the agreed price. Generally, economists might issue that it is easier to give secret rebates or discounts in the freight rather than the passenger market again, because there are fewer customers and the information is therefore less likely to leak.

Price Leadership

Another possible theory of oligopoly is price leadership, which can be regarded as another form of collusion. Price leadership can occur, when the biggest or most powerful firm can raise or reduce prices knowing that the other firms will be obliged to follow. The biggest firm in the industry is not always the most powerful one, since sometimes the transport organisation may be part of a much bigger organisation, which has more power. If there is collusion, then the industry can act in a similar way to monopoly (i.e. by restricting the output it can raise the price). Sometimes as with IATA, it could be claimed that IATA was the price leader.

Kinked Demand Curve

The Kinked demand curve was devised by Paul Sweezy (1910–2004) in the interwar years (1939) to show why there was comparative price stability in many oligopoly markets. It should be realised that the period concerned was one in which inflation was generally low and that there was considerable unused capacity because of high unemployment. The assumptions of the kinked demand curve were that firms would be unwilling to reduce their prices on the assumption that other firms would follow suit and with a comparatively inelastic demand that all firms would therefore be less profitable. However, a firm would be unwilling to raise its price since if other firms did not follow suit it would lose a large part of its total market. If oligopoly is one where the product or service is homogenous in the eyes of the consumer, then a rise in price would lead to a complete lack of demand.

The Kinked demand curve was less likely to be true in the 1970s where most transport organisations would have been under pressure to raise prices to reflect the increase of costs following the Oil Petroleum Export Countries (OPEC) price rises in 1973 and 1974 and immediately following the Iranian revolution in 1979. It would have been less likely to apply in the 1950s and 1960s where in the UK as in many Western countries full employment would have meant that there were broadly good load factors. Therefore, organisations might have been able to raise prices without a fear that other companies with excess capacity would have undercut them considerably. Currently in the UK where the target rate of inflation is 2%, this is less likely to apply.

The theory of games suggests that firms will not only look at their own decisions, but also the possible reactions to their moves. A small bus company might be reluctant to indulge in fare cutting against a larger rival on the assumption that if they reduce their fares that the bigger firm might more than match this. The larger organisation even if it does not cover its variable costs in the short run because of massive resources might undercut the smaller firm and thus lead to larger profits in the longer run. Economists can observe game theory in the UK transport industry, and see how the railways, coaches and airlines compete on the Inter-City market within the UK. When the former British Rail introduced its cards for students in an attempt to gain the greater demand from lower income people, some of the coach companies immediately followed with a similar scheme. Similarly, when the airlines introduced their shuttle services former British Rail sometimes made strenuous efforts to reduce the timings of some of its journeys and to have more stops at a short distance outside the towns, for example at Watford about 10 miles out of London in an attempt to reduce accessibility time and thus overall journey time compared with its airline rivals. Whilst the theory of games gives reasonable predictions for an individual industry or intermodal rivalry in a given case, it gives very little guidance as to what will happen under oligopolists' conditions as a whole. It therefore requires considerably more research than the more straightforward kinked demand theory.

Charging Below Variable Costs

Whilst economists usually stress that firms will not knowingly charge below variable costs, this is not always true. Outside the transport industry, firms have been known to set up fighting companies (i.e. companies, which are nominally independent but in practice controlled by the parent company), which have reduced prices below that of their rivals, even to the extent of going below variable costs and have then vanished once the competition from smaller companies has been eliminated. It was claimed in the early 1980s by Freddie Laker who formed Laker Airways that the larger airlines engaged in similar predatory pricing (i.e. charging below costs in order to eliminate their rivals). It is very difficult to determine the truth or otherwise of these assertions though it has also been claimed that this was true in the 1920s when the bigger bus companies, contrary to general opinion, frequently undercut their rivals rather than the other way around. Partly for this reason, therefore some economists would not only wish to have an upper maximum price to limit oligopoly or monopoly behaviour but might also wish to impose minimum prices to avoid such predatory pricing. Therefore, the EU had proposed there should be a forked tariff system (i.e. imposing both minimum and maximum prices for international road haulage). Usually transport economists would associate maximum price controls with monopoly and minimum price controls with wasteful competition.

Monopoly – What Is the Definition of Monopoly?

The economists' definition of monopoly is that of a good or service in which there is no substitute in the eyes of the consumer. It is extremely rare to find an example of complete monopoly, though a rural bus service may well come into this category for non-vehicle owners. Organisations can be monopolies in some services but not in others. British Railways was certainly not a monopolist on most of its Inter City Network. It competed with coaches at the lower end of the price range especially on shorter journeys, and with airlines for higher price travel, particularly on longer journeys such as those of London to Glasgow. On some of these journeys, it also competed with the private motorists.

On the other hand, it was at least in the short term, a monopoly for longer distance commuter traffic such as that from Peterborough to London. The same also applies to the privatised network. The journey from Peterborough to London Kings Cross is extremely fast and therefore the operator will have some monopoly power. The 1980 Transport Act was expected by some commentators, including the late Professor John Hibbs (1925 to 2015) to alter this as coach firms could more easily enter the London commuter and other commuter markets. In practice, however, the competition was generally short lived for the market as commuters appreciated the greater frequency of British Rail and also the better time keeping of rail rather than coach services within the London commuter area. However, where coach commuter services travel directly to areas not well served by rail, they have carried many passengers.

Concentration Ratios – Measures of Monopoly Power

Monopoly can be measured by concentration ratios (i.e. the proportions of the top three or four firms' assets sales, revenue, of the whole industry). These figures may well give differing ratios, looking at revenue, then organisations, which have high fares or charges will be a larger per cent of the market than firms charging less. If economists look at the number of ton-miles or passenger miles, then they will get different results.

Cruise operators, mainly catering for the wealthy, will charge considerably more than those which have much lower standards of accommodation and will therefore appear to be a bigger part of the market judging by passenger kilometres alone. This may well be misleading. Also in the road passenger industry one large taxi firm charging high fares would appear to be a large proportion of the total market judging solely by the revenue. Looking at the number of passenger kilometres will give a different pattern.

If economists measure concentration by volume of employment, then a labour-intensive service, such as a road haulage firm, would appear to be proportionately bigger than a capital-intensive one. Similarly, the Docklands Railway in London does not have drivers and most of the fares will be paid by machines, then it will appear smaller than a bus company, where the bus fares are collected on the older buses.

Local Monopolies

Organisations are sometimes purely local ones; therefore, a low concentration ratio could still mean that there were monopolies in particular towns or regions. Until the changes in the 1980 and 1985 Transport Acts, Passenger Transport Executives could have had a monopoly on bus services in a particular conurbation even though they would not have formed a large percentage of the national revenue or passenger kilometres. Where firms compete on an international basis, it may be difficult to find figures in terms of value, which are on a comparable basis because of fluctuations in exchange rates. Following the British decision to exit from the EU (i.e. Brexit), the pound fell considerably against the dollar. If transport economists have looked solely at dollars, the British companies would look much smaller than they had been previously even if they still carried out the same number of journeys. This would be important in the aviation industry.

Sometimes because of confidentiality, it may be impossible to find a comparable figure. Concentration ratio figures may be available on an industry basis but may not be helpful when different modes of transport compete with each other.

Other Methods of Measuring Monopoly Power

Supernormal profits over a long period also show the existence of monopoly, since supernormal profits cannot be made under conditions of perfect competition in the long term. However, the absence of supernormal profits would not necessarily indicate that there was not a degree of monopoly. For example, the 1972 investigations into the various English Channel Shipping operators by the then Monopoly Commission showed that supernormal profits were not being made by the operators. This may have been because of comparatively poor pricing methods and efficiency of the then existing operators.

Factors Affecting the Degree of Competition

Terminal operators including airport operators are more likely to be monopolies than vehicle operators. This arises because terminals usually have a much longer life than vehicles and because often there are a limited number of potential sites, which could be used. There is more reason to believe that wasteful competition would take place, if there were too many terminals. Nevertheless, international competition may occur between airports or seaports. One of the arguments in support of a third London airport during the 1970s was that without this further capacity, airlines might use continental airports and Britain would lose revenue. The EU has suggested that there should be an overall ports policy, partly because of fears of excess competition. To date, little has been done about this suggestion.

One of the worries expressed over the privatisation of the then British Airports Authority was that it would not necessarily lead to any greater degree of competition. It was almost inconceivable that any major airport development could take place, for example in the South East and where the British Airports Authority controlled Heathrow, Stansted and Gatwick and therefore BAA had some monopoly power. In 2008, the government announced that Gatwick would be sold.

Economic Fashions and Competition

Government attitudes towards competition and monopoly seem to vary, partly according to economic fashions. The Victorians (1837–1901) placed great emphasis on competition. Sometimes preoccupation with competition was excessive since there were two rail bridges in Rochester across the Medway which would not have been justified by the volume of traffic and was wasteful. However, Sir Robert Peel, the Conservative prime minister, was more inclined to interfere with the laissez faire doctrine than his predecessor the Duke of Wellington.

Government was less inclined to view competition as important in the 1920s and 1930s. This was partly because the government had become used to having some control over the transport industry during the First World War where it was vitally important to ensure that transport was geared to the war. This was particularly true of the railways, which in 1923 formed the great four, i.e. the LNER, the LMS, the GWR and the Southern Railway. The 1930s was one of economic depression and therefore government was often inclined to try to deter foreign competition. The 1960s was often dubbed a "Big is beautiful" era and there was encouragement for large-scale firms in both the public and the private sectors.

In the 1980s, there was a reversal towards the "Small is beautiful" philosophy of the late E.F. Schumacher (1911–1977). In the transport industry, this may account for the Conservative Government wishing to split the former National Bus Company into a variety of smaller units after its privatisation during 1987 and 1988. Since then the bus industry has been dominated by the large companies and the same is true of the privatised railways. Transport economists will observe that Virgin is one of the major companies in the rail industry and is very important in other parts of the transport sectors. Similarly, both FirstGroup and Arriva have both interests in rail and buses.

A reason for the existence of monopoly is the high costs of entry into a market. This could apply to the supertanker market where the cost of individual ships can be very high. It would equally be true of the Civil Aviation Industry. However, a degree of competition in the supertanker market is still likely to occur because the oil companies themselves may buy ships.

Is Competition Desirable or Undesirable?

The static model of competition, which has influenced successive governments, in the UK and elsewhere, implies that monopoly is undesirable since monopolies may have an incentive to reduce supply and increase price. This is in contrast to perfect competition. This argument ignores the potential economies of scale. Some economists suggest it is sensible to look at different modes rather than to assume that it is desirable across modes. There is relatively little evidence of economies of scale in road passenger or freight industries. Therefore, economists might reasonably view restrictions on competition on road freight or road passenger as undesirable. A second argument, which is often advanced against monopoly, is that it may lead to management slack, as the great advantage of monopoly is a comfortable life. X-inefficiency is the difference between efficient behaviour of businesses assumed or implied by many right-wing economists. They assume that observed behaviour in practice is caused by a lack of competitive pressure. The concepts of X-inefficiency were introduced by Harvey Leibenstein 1922–1994. This advantage is to the monopolist and not the consumer. The evidence here may well be conflicting. Certainly, some large-scale public operators, such as London Transport from 1933 onwards, were relatively slow to introduce innovations. On the other hand, it is sometimes argued that a monopolist will have the necessary economic security to undertake research and development, which might not apply in a more competitive situation. The arguments regarding other industries, for example by John Jewkes 1902–1988 and the late J.K. Galbraith 1908–2006 are somewhat conflicting.

A further argument against monopoly is that it restricts choice. However, both IATA at the time of the standard airfares and shipping companies would suggest that in practice instead of bunching around particular times or restrictions to the number of services, in fact, they offer more choice. There is a limited amount of evidence to support this assertion from other modes of transport. It has been reported that in the North East of England since the 1985 Transport Act road passenger operators have chosen similar times for their buses, which may restrict rather than extend choice for the consumer.

There were suggestions that to give more railway competition, that one authority could provide "The Way" whilst separate undertakings provide the rolling stock, etc. It has been stated erroneously that the EU requires this when what the EU requires is that there should be transparency of costing (i.e. a lower price than necessary should not be given by the track authority in order to subsidise the railway operators). Since the formation of Railtrack and then Network Rail the track, authority and operations are separate except in the case of the Isle of Wight railway. Many commentators have felt that the separation of track and operations leads to problems and in particular has led to the problem of costing and lack of interest in electrification. Sometimes, as with the main traffic between the UK and the USA in air transport, the unit of output itself requires a large sum of money. This is equally true of shipping and oil tankers. Nevertheless, one can find small firms (e.g. the service provided by air taxis).

Statutory Monopoly

Sometimes governments have given some degree of statutory monopoly power to undertakings. This is nothing new. It existed under the Navigation Acts from 1651 onwards and into the eighteenth century. They restricted imports to the then American colonies to either British ships or those of the country from which the imports or exports were going.

Official attitudes of government towards monopoly vary. In the UK, there was no formal legislation against monopoly until the 1948 Monopoly Act. The British government has not taken the view that monopoly per se is undesirable but that each case should be treated on its merits (The Competition Act 1998 and the Enterprise Act 2002). There have been subsequent acts in 1973, 1980 and 1998 and the enterprise Act 2002, though the British governments have not necessarily taken the view that monopolies are to be prevented.

There have been relatively few investigations by the Monopoly Commission (later the Competition Commission and now the Competition and Market authority) of transport undertakings, though one was of the various Channel Ferry Operators in 1972. On the other hand, in the USA, the Sherman Act 1890 had taken the view that monopolies are not advisable. In practice, monopolies are frequently tolerated. Nevertheless, in the late 1970s and 1980s there were some signs that US governments had taken a different viewpoint, towards shipping conferences from other governments. The British government had indirectly encouraged regional monopolies with the road passenger licensing system, which existed up to the 1980 Transport Act. In road haulage, however, the licensing system does not seem to have encouraged larger firms, even though until 1968, it was difficult for new firms to enter the industry. There has, however, always been competition from own account transport.

The former shipping conferences were referred to as cartels because they influenced both the output and the price of their members. The shipping conferences denied that they were cartels pointing out that there were no restrictions on other firms entering the industry. The West European shipping industry in the late 1970s and early 1980s had been subject to increased competition from the Eastern European countries whose objectives were not solely profit. The then Eastern European countries often had balance of payments problems and one way around this was to ensure that they gained trade in the shipping industry even if it were not profitable so that they obtained hard currency.

In air transport, until the breakup of the IATA agreement on passenger fares, there was little competition, except on a non-price basis for scheduled flights. However, the growth of the chartered flight market meant that it had become increasingly difficult to segregate these markets. The air industry indicates one of the problems in trying to classify transport, since scheduled flights increasingly faced competition from chartered flights and relevant data was not always easy to obtain. The case of air transport is an interesting one since there was a considerable amount of competition on chartered flights, but relatively little price competition for scheduled flights. Often on domestic flights, and on some of the shorter international flights, they face competition from other modes of transport. The advent of the 'Sky train' operated by Laker Airways, and various similar airline schemes meant that, on the North Atlantic run, there was more competition. On domestic flights between European countries, however, prices had remained high. The advent of budget carriers with their no-frills airlines coupled with the growth of internet bookings meant that information became better. Budget airlines could offer very cheap flights, which often stimulated a great deal of travel. There have, however, been concerns since the air transport industry, has been one of the most rapidly growing sources of pollution. However, higher fuel prices in 2008 and the credit crunch means that this growth may not continue. The unexpected collapse of oil prices in 2015 makes it even more difficult to predict what will happen. In most countries, until recently, though not in the United States of America, there was only one major railway company. However, this did not always confirm to the economists' definition of monopoly.

Causes of Monopoly

The most likely cause of monopoly is restrictions on entry either through the government, with the licensing systems for road passenger or road freight or because of lack of access to certain facilities. Newer operators have sometimes complained that it is difficult to obtain permission to fly from Heathrow and this debars them from competing with the major airlines.

A monopoly may exist because of its greater efficiency. According to monopoly studies, this has been true of many industries though it is difficult to find an example of this within the transport industry.

A third reason for monopoly is the influence of government. The Conservative government set up London Transport as a monopoly of both bus and Underground lines in 1933. This had virtually remained unchanged until 1987 and 1988 when tendering was allowed for a wide number of bus services. Whereas most bus services have declined since the 1980 and 1985 Transport Acts this has not been true of the London area. This, however, may be because of the congestion charge in London.

A fourth reason for monopoly is that firms may take over or merge with other companies. Hence, government influence may be a contributory factor. Until 1923, there were about 120 independent railway companies. Subsequently, they were regionally grouped into LNER, GWR, LMS and SR.

A fifth possible reason for monopoly is the difficulty of finding suitable outlets for selling its services. This can be found in some industries but is less untrue of the transport industry where there are many forwarding agents and travel agents who would usually be receptive to a new company or new services.

Workable Competition

Economists now use the word workable competition to mean that firms could enter the market, even if in practice they do not so that the threat of this may help keep down prices.

Collusion

Collusion is most likely to take place, when the costs of different transport undertakings are broadly similar. It is therefore unsurprising that there has been collusion in the air transport industry namely the IATA price agreements as many airlines have very similar aircraft and therefore similar operating costs. With similar costs, it becomes easier to determine a price which would give reasonable returns to all firms and where no oligopolist is tempted, perhaps because of these, to reduce price. However, where cost conditions vary, as in the shipping market with competition between hovercraft, hydrofoils, catamarans and conventional ships, it is more difficult to see how collusion could occur.

Collusion is also more likely to take place, if it is possible to keep out newcomers from the market. This may apply within domestic transport because of the limitations on competition imposed by the government. In international transport, it may also apply because countries may be reluctant to allow foreign competition as with air transport.

Another possibility is price leadership, which may be regarded as another form of collusion. Under these circumstances, the biggest organisation may be able to raise or reduce prices knowing that the other organisations will be obliged to follow.

Collusion may take the form of an agreement on price, although not on quality of service. Under these circumstances as, for example, with a great deal of the IATA scheduled services until the early 1970s, competition will then take place on a non-price basis. This led to newer aircraft and to a greater quality of service then would occur under perfect competition. Any airline would be perhaps willing to do this on the basis that to introduce a newer faster aircraft would give them an increase in their market. Another possible method of collusion is pooling of revenue, as occurred between British Airways and Air France. Therefore, the opposite conclusion would hold. The airlines would have no great incentive to innovate, as this would usually mean higher costs without a proportionate increase in revenue.

CHAPTER 6
Pricing Policy

There are a variety of different methods of pricing in both passenger and freight transport. Thus, it is extremely important to understand the economic theory behind the different pricing schemes.

Government Control and the Railways

Pricing policy depends upon the organisation's objectives. The nationalised industries objectives were given by the relevant statutes (e.g. public railways in the 1974 Railway Act as well as White Papers such as those in 1961, 1967 and 1978 and by the overall government economic policy). The objectives of private sector firms will be partly stated by the memorandum of the association and more specifically by the board of directors and the individual departments within the firm. Since the 1993 Railway Act, the private companies have had their pricing policy determined by the Rail Regulator, which has been renamed the rail and road regulator. The regulator has often specified maximum fare increases to be limited to the rate of inflation plus or minus a specified amount. The Office of Rail Regulation was established in 2004 under the Railways and Transport Safety Act 2003. This is now the Office of Rail and Road regulation (ORR). Sometimes, as with the domestic services to and from the Channel Tunnel from 2009, there has been a premium rate of inflation plus 3% for journeys to and from St. Pancras, but inflation plus 1% for other journeys. For other modes of transport, such as the airlines, there are usually no limitations, so prices charged will aim to make the operator gain the largest profits.

With privatisation of the rail network, the private rail companies sometimes introduced maximum price rises based on the rate of inflation and to some extent as well upon the quality of service provided. This had been below the rate of inflation but in 2007 was inflation plus 2% partly to allow for investment. Pricing policy will depend upon the type of market.

Pricing in Tramp Shipping

In tramp shipping, which approximates to perfect competition, there is little choice about pricing policy, since the firm has to be a price taker. Therefore, prices will reflect costs allowing for what the economist describes as normal profit. However, in oligopoly, pricing will have to take account of a rival's reactions. This can be seen on routes where there are a few operators or sometimes across modes of transport where the Channel Tunnel will have to take account of the shipping companies prices and vice versa. In a pure monopoly, which is comparatively rare, the operator will probably use price discrimination. Price discrimination means a different set of charges, which are unrelated to differences in costs. Different fares charged for people over 60 will therefore be an example of price discrimination, whereas different rates in the peak are not.

Restraints on Pricing Policy

Organisations may be limited in their pricing policy, since they may not wish to attract the attention of government bodies dealing with monopoly such as the Competition and Market Authority (CMA). This is particularly likely to be true of the road passenger industry since the 1985 Transport Act.

Standard Rate per Mile or Kilometre

This has been a common form of pricing for passenger traffic. British Rail in the early 1960s for journeys up to 200 miles, had a standard rate of 3 old pence (1.25 new pence) per mile. With the extensive rail network, which British Rail then operated, it had the advantage to the operator of being a simple system providing the distance was known. There were, however, sometimes problems when there were several routes serving the same destination as from Waterloo or Paddington to Exeter. Either therefore, two separate fares would have to be charged or more usually, an artificial mileage was agreed.

The advantage of the standard rate per mile was that it was generally acceptable to public opinion; however, the disadvantage was that it took no account of the different costs to the undertaking of providing mainline, suburban or rural passenger services. The cost differences between these were often very substantial. It also took no account of potential passenger demand or competition. Therefore, it was easy for road passenger operators to pick up the longer distance mainline traffic which the railways were often well suited to whilst the former British Rail carried a disproportionate amount of rural traffic. It could also lead to overcrowding on some routes at certain times of day where higher fares would have gained the railways more money and might have led to less overcrowding if people moved to other times of the day. Sometimes, the railways might have been able to gain more revenue if, as is possible, the price elasticity of demand is greater than one for some social passenger traffic.

Tapering Fares and Charges

A variation on standard fares is tapering fares where there is a proportionately higher charge for short rather than long journeys. There could be a passenger fare of 50p per mile for journeys up to 6 miles and then the next fares might be £3.25, £3.50, £3.75, etc. The logic of a tapering fare is that passengers would be more likely to be deterred by high fares especially for shopping for longer distances. It is less economically justifiable for business journeys where there may be fewer less option about taking the journeys. Some research particularly by W. Tyson from Liverpool University has stated that bus undertakings may have raised the very short distance fares by too great an amount and that therefore passengers may have substituted walking for travelling for very short journeys.

These fares have often been issued by UK bus companies. The logic is that a standard charge per kilometre or mile would deter long distance passengers and also that there is a boarding charge since bus operators' costs, have two elements the acceleration and deceleration times and also the time during which the passenger holds up the vehicle which are incurred irrespective of the distance travelled. The other element is the distance travelled which means that more fuel is being used. Tapered fares can be coarse (i.e. only a few fares are charged). They can be fine which means that there is a multiplicity of stages. Coarse fares for road passenger undertakings have the disadvantage that passengers may be prepared to walk short distances to reduce their fares. Fine fares on the other hand have the disadvantage of being complex and they slow down boarding time unless a system of computer checks can be introduced on the buses. If tickets can be issued off the bus, then this does not matter too much. The Underground in London has usually had in the past a series of ticket machines for different fares so that a system of coarse fares was desirable so that there were not too many different types of machines. Modern methods of issuing tickets have made this less of a problem, but there are still in London a series of zonal fares. The railways by contrast, apart from London and some other conurbations often have a great many ordinary and discounted fares, which the passenger can be charged. Even conductor guards now have ticket machines, which can cope with a wide variety of different fares.

Tapering Fares and the Airlines

The logic of tapering fares will also apply to airlines where the cost of short hauls is much greater than that of longer hauls. This is shown by the relevant data for the former British European Airlines (BEA) and which always had much higher costs per kilometre travelled than the former British Overseas Airways Corporation (BOAC). There is less obvious difference in costs between long and short journeys, where there are few stops as in the coach or mainline railway services.

Tapering Charges and Freight Traffic

Tapering charges also apply to most freight traffic where there is a high initial cost of handling which will occur irrespective of the distance travelled. The use of unit loads has, however, often reduced handling costs and therefore there is slightly less need for a very strong taper. The container revolution has also reduced the scope for the traditional method of charging what the traffic will bear.

Flat Rate Fares on Buses

Transport economists can find examples of this in both passenger and freight transport. Many European countries adopt this system on their bus routes as in Paris or Rome. The advantage of the flat fare system is that it is easy to sell the tickets in books (carnets) and this is done by many European bus organisations, for example in Amsterdam, Rome and Paris. London has more recently given carnets as well and Oyster cards which are paid for in advance. These tickets are often sold by a wide variety of retail outlets, which gives further publicity to the undertaking as well as reducing the administrative costs to the undertaking. Using retailers however does mean that commission has to be paid to them. The use of carnets is not however confined to road passenger undertakings, since it is also used widely on many metro systems.

Post Office Charges

The Post Office system of charging the same rates for letters and parcels of the same weight irrespective of distance may be regarded as a variation of the flat rate system. It leads to the cross subsidisation of the longer distance rural postal services (e.g. from Cornwall to the North of Scotland by urban post). Such a system would be very vulnerable, if the Post Office did not possess a monopoly of such traffic. This is one of the concerns of the Post Office especially now it has been privatised. There is concern that competitors will concentrate on certain towns (i.e. cherry picking within a large area) when there will be larger scale deliveries, which are obviously cheaper.

Zonal Charges

The refinement of the flat rate system is a zonal rate, which is both used for parcel traffic by many haulage undertakings and on some passenger undertakings.

The flat rate system has the advantage of administrative simplicity though it takes no account of the costs of providing such a service. In 1981, London Transport later London Regional Transport introduced this. The flat fare was the system on most suburban routes. The disadvantage of changing the same fare for all distances was that longer distance journey users might have been willing to pay much more than 25p the then current fare. Longer distance journey users have a considerable degree of consumer surplus. Shorter distance passengers may, however, have been deterred from using the service. There have since been a series of modifications to the policy and in 1988, the flat fare in suburban routes was 40p except in the morning peak when the fare was 45p. For short hops, up to about 2 kilometres, it was 30p and 35p respectively. In 2009, the flat fare for the suburban journeys was higher in the central London area than in the suburban area. One of the advantages of a flat rate system especially for buses is that it speeds up the loading of buses and it is easy to issue the pre-booked tickets. However, there are several methods of speeding up loading buses including issuing travel cards, which do not penalise people who make multi stage journeys. The use of a check system as used on London

Transport as part of their undertaking in 1987–1988, which checked tickets for validity, also meant that there was more scope for different types of pricing. The use of zonal cards helped to overcome the problems of multi stage tickets. Transport for London now has a system of zonal cards, which cover rail, bus and Underground in 6 main zones, although sometimes a station may be in two different zones. The ticket machines can check the validity of the tickets and this includes tickets issued from outside the London area. The system also can check season tickets, which helps to speed up the flow of passengers. It was hoped that the Oyster smart card system could be developed to include many rail routes especially those starting or finishing in London. The London Overground also uses this system. The oyster system has been extended.

Price Discrimination

This means differences in prices, which are not accounted for by differences in costs. Transport economists can observe several current methods used by the UK rail companies. These include family cards, senior rail cards which are used by anybody over 60 and cards which can be used by anyone under 24, as well as full time students of any age. There is also a disabled railcard, which can be used by both the disabled person and an escort at any time. For all of these cards, users pay a flat rate per year, but then pay a reduced rate on most rail travel. There is also a Network South East card, which gives a discount of 30% on off peak journeys over £13 during the week and on all journeys at weekends and on public holidays.

Price Discrimination Used and Its Obstacles

Price discrimination may be used for either profit maximising reasons or social objectives. Price discrimination when used for profit maximising aims to reduce consumer surplus. If monopolists have perfect information and could segregate the markets without cost, then consumer surplus would be entirely eliminated. In practice, this is obviously not possible. Price discrimination will be used when the price elasticity of demand is in different markets, but it must always be possible to segregate the markets. The widespread availability of photo-cards makes it much easier to do this currently than in previous decades; though somewhat surprisingly the family card no longer uses photo-cards, neither does Network South East. Price discrimination has been used for a mixture of social and commercial reasons. Nearly all transport undertakings have a special fare for children, which especially at peak times may not be justified solely on commercial grounds. In France, there have often been cheaper fares for older people and particularly for veterans of the armed forces for social reasons.

Price discrimination may go to great lengths especially on airlines where one report suggested that on one flight, thirty-three different types of fares could be used. The internet has made price discrimination much easier, since if people wish to book and there are still some seats left the airline operator can decide on the range of prices to maximise the yield.

Price discrimination has not been confined to the passenger sector. The UK railways for a long while had a very complex system of charging according to what the traffic would bear. Under this system, live batteries were charged a different rate from dead batteries though there was no difference in the cost of transporting them. The use of containerisation, however, has made it much more difficult to segregate the market and a greater emphasis on handling speeds has meant for example that in short sea ferry routes it is usual to charge for containers for lorries, by length rather than by any other system.

Season Tickets

These have been used both by the privatised railway companies and its predecessors. Bus undertakings are increasingly using them though sometimes with variations, for example allowing passengers far more choice of routes than traditional season tickets, which specified the points between which the season ticket could be used. Season tickets have the advantage of administratively low costs, they give a clear indication of overall demand though do not indicate the exact timings of journeys. They also improve operator's cash flow especially the long-term tickets such as those issued for a year. From the passengers' viewpoint, they avoid the need to queue except at long intervals and have the advantage that marginal journeys can be made at no additional cost. This may be especially helpful to the commuter who uses the season ticket mainly for journeys to and from London or other conurbations for work purposes on Monday to Friday but might wish to make social visits at the weekend. Even the queuing at intervals is no longer true, since they can be sent through the post. The railways have tried to encourage the use of annual tickets by giving special concessions to the annual ticket holder. The gold card gives the holders cheaper rates at weekends.

When there are high rates of inflation, this will speed up cash flow to the rail companies. Customers gain as it gives them certainty about fares.

Fares by Class of Journey

The UK railway companies have two main classes (i.e. First and Standard Class). Whilst differences in these fares partly represent differences in cost, they also take into account the different willingness to pay. Former British Rail in the 1980s reintroduced Pullman Class. This is partly because whilst British Rail faced competition from the coach services following the 1985 Transport Acts where the competition was mainly based on fares, and where the price competition was particularly important from the student coach market with National Express introducing a student card system with the slogan 'It is much cheaper by far than British Rail'. However, the Pullman class was reintroduced to compete at the other end of the market where the emphasis is not so much on price but on quality of service.

The short haul low cost airlines in 2007 took away some traffic from the rail companies on the London – Glasgow route and to a lesser extent the London – Manchester route especially during the upgrading of the West Coast main line which often meant that journey time would have been considerably longer. Virgin rail services faced competition internally in 2004 from the St. Pancras – Manchester route as the now disbanded Strategic Rail Authority gave permission for the then Midland Main Line service to be extended to Manchester. Whilst taking longer than the usual service, the prospect of a route without changing may have enticed some passengers away.

Whilst there is a wide variety of different airfares, the underlying strategy reflects that there are two main distinct markets, social travel (including tourism and visiting friends and relatives) and the business market. Until the early 1970s, the chartered market had grown at a much faster rate than the scheduled airlines which had catered mainly for the business market. This situation had arisen partly because of the IATA system, which had regulated fares on scheduled flights but not on chartered flights. This led to waste since there were relatively low load factors on scheduled flights because of the high fares charged. The airlines frequently use an Apex system, which means that fares are cheaper if passengers order them in advance. The logic here is that social travellers will usually know in advance when they are travelling, but business executives wish to have flexibility about when they are travelling. The railways in UK have also called some of their tickets Apex for similar reasons.

Peak and Off-Peak Fares

The railways have always used this system. British Rail operators have conventionally used this with commuter traffic to have higher fares for journeys beginning before 9:00am or 9:30am. They have sometimes imposed restrictions on the return evening peak between 5:00pm and 6:00pm, though more recently this has not often been applied. For business journeys, however, the peak is likely to be at different times usually slightly later than the journeys to and from work. British Rail operators often use the off-peak day ticket, which is issued for journeys up to about 50 miles. The off-peak day ticket is made cheaper in the South East of England by the use of the network card in the South East where the journeys have to be made after 10:00am. though there are no restrictions at weekends or on Bank holidays. In the nineteenth century the railways used to issue a workman's ticket which was valid for journeys made very early in the morning. Currently this has occasionally been reintroduced with the "early bird" ticket.

For longer journeys, intercity saver tickets are generally available which give substantial discounts approximately one third over conventional returns. British Rail used a practice which is used on other European railways mostly France of having White Saver Days and Blue Saver Days. Blue Savers are cheaper than White Savers and generally apply to most days with the exception of Friday, which is a peak day both for business and social travel. The savers are made cheaper when used in conjunction with the other rail cards, such as Young Persons Card, and Family and Friends Railcard. There have been occasional restrictions on the use of such cards, such as for journeys for example from Kings Cross between 6:30am and 9:30am and they could not be used until 6:00pm or later on journeys from Kings Cross.

The airlines have also made use of off-peak fares. For holiday journeys the peak is likely to be the summer months and so, therefore, lower amounts are generally charged in the off-peak months. The airlines have often introduced a limitation to reduce the number of business travellers making use of lower price fares by insisting that travellers have to stay over a weekend.

Payments for Higher Quality of Service

This is the payment for higher speeds, which is one of the reasons why certain types of transport, such as taxis can generally charge much higher fares than bus operators. However, increasingly operators have realised the advantages of operating different types of services especially as in the UK sector there have been fewer inhibitions on pricing policy. The Rapide Coach Service with its toilet and video service meant that prices could be somewhat higher on longer journeys than would be the case with the conventional service. Prices, however, are kept to a reasonable level partly because of the high load factors. Other operators in other countries have often used this method. Railways in the former West Germany traditionally charged more for fast rather than slow trains and this has continued with a premium paid for the ICE trains. This has happened on occasions for example in the 1970s with British Rail where British Rail used the spare capacity on an East Coast route to charge lower fares than on a conventional main route whilst still giving a time advantage over coach operators. Airlines charge more for some non-stop services with better seating but this reflects the higher quality of service as well as speed.

Standby Fares

These have been used mainly by airlines but also sometimes by coach companies. The marginal cost of additional passengers if there is spare capacity is extremely low. Therefore, additional passengers, if they can be obtained without losing fares from existing passengers, are very welcome. The basis of the standby fare is that there is a lower fare charge though seats are not guaranteed. It relies on a market of passengers, who do not mind too much if they catch a later flight or a later coach journey. Therefore, the price differences must at least be significantly greater than the financial costs of travel to and from the coach or air terminal as well as some allowance being made for the value of time to the travel. One potential problem with the system is that if there are regular travellers particularly business travellers who become aware that there are always available seats on the service that they become aware of this and will only pay the standby fare rather than the regular fare. This is generally likely to be less of a problem for the coach owner where there are less likely to be regular travellers.

Marginal Cost Pricing

Many economists have advocated marginal cost pricing, because it leads to a Pareto optimum subject to a number of constraints. The term Pareto optimum means that it is impossible to make any consumers better off without making others worse off. In transport, there are practical problems arising with the use of marginal cost. Do economists mean the marginal cost of the individual, the railway carriage, the train or the service as a whole? A particular problem arises in transport because generally the unit of carriage is not the same as the unit of demand.

There are also problems about whether operators should use short run or long run marginal cost pricing. If economists take the Isle of Wight line as an example, in the short run the only costs will be that of driver plus station staff. Charging short run marginal costs would therefore mean covering these costs. This might be sensible and would be helpful even if it covers the service costs providing there are no major repairs, etc. It would thus make the best use of existing assets. However, in the long run, if the stock needs replacing or there has to be track maintenance charging short run marginal costs, it does not give any criteria as to whether passengers would be prepared to pay for this. The use of long run marginal cost of pricing taking account of these costs, would give better criteria for investment but would not make the best use of the existing assets.

In practice, because the unit of carriage is not the same as the unit of demand, there is bound to be some averaging of cost. If we took marginal cost pricing too literally and considered a train with a potential capacity of 500 passengers, it would mean that the first passenger would pay a very large sum of money whilst the remaining 499 would pay a very small amount representing additional wear and tear and slightly greater fuel consumption.

It is possible to overcome the short run marginal costs and long run marginal costs dilemma by charging short run marginal costs but guessing, possibly using questionnaires whether passengers would be willing to pay the long run marginal costs. The use of standby fares on both coaches and aircraft at reduced fares is an example of the principle of marginal cost pricing. Marginal cost pricing only leads to the Pareto optimum and the best allocation of resources if other sectors similarly charge marginal cost. If British Rail had adopted a marginal cost pricing policy as stated in the 1967 White Paper and also by the EU, but road hauliers charged considerably more than this, then the railways would be carrying more traffic than under a Pareto optimum system.

Another problem is whether private marginal costs or marginal social costs should be charged. It is unlikely that these will be equal, particularly in transport, because of the externalities such as pollution, noise and accidents. If the railways charge the marginal social costs on rural routes where people might otherwise be unemployed this would be desirable but would almost certainly lead to financial losses, which would certainly be unacceptable to the present government. Charging private marginal costs might, however, overstate the true costs of the railway system in such rural areas.

What Happens If the Marginal Cost Curve Is Declining?

If the marginal cost curve is declining, charging marginal costs will lead to losses. This is because the marginal cost is lower than the average cost. Such losses could lead to distortion in prices elsewhere since this would have to come out of taxation. This problem can be overcome by using price discrimination and using marginal cost as a minimum or alternatively by using a two or multi-tier tariff system. Two or multi-tier price systems will mean that the passenger or freight customer will pay a fixed amount and then a further amount according to the variety of usage. This occurs with a great deal of the service sector of public utilities in electricity, gas or telephones where the user pays a rental and an additional amount reflecting his use. This system is not generally appropriate to public passenger transport though to the use of student, senior and family cards on the railways. The use of a ticket which is paid for people over 60 with the senior rail card and then allowing them lower price elsewhere, can be regarded as an example. For freight, it is clearly more applicable, although it may not be used in the form described. There are often lower rates according to the greater usage.

Marginal Cost Pricing and X Inefficiency

Marginal cost pricing will not make the best use of resources if there is X inefficiency. X inefficiency was a concept publicised by H. Leibenstein. It refers to the difference between the actual cost of an organisation including transport organisations and the lowest costs which could be obtained if firms optimised their allocation of resources. One of the reasons between the differences are that managers and other employees may well have objectives, which differ from those of the firm. In the transport industry, it is very difficult to measure productivity and, because of the scattered nature of the industry, extremely difficult to check that people and equipment are working with maximum efficiency. This is obvious to anyone who has waited patiently or impatiently for a variety of different transport modes. The perfect competition model assumes that firms cannot survive if they are inefficient. In practice, this only applies if there are not any market imperfections, such as the difference between perceived and actual costs to entry in the transport industry. The importance of X inefficiency is that it means that prices do not necessarily reflect resource costs. Therefore, the marginal cost of the organisation may not reflect the resource costs.

Marginal Cost Pricing and Income Distribution

Marginal cost pricing may not be adopted in practice because of income distribution reasons. In the UK, people over 63 have free off-peak bus fares. In the early 2000's the government made it compulsory for local authorities to issue a free pass for the over 60's which gave half price on the local buses. This concept has now been extended so that from 2006, local bus services have been free for over 60's for local off-peak bus journeys and from 2008 for all off peak bus journeys (see the Freedom Pass). This minimum age has slowly been increased. This is below the marginal cost but may be thought to be desirable on income distribution reasons. In other countries too, prices may be held down for certain categories (e.g. in France for large families, and war veterans for the same reasons).

Fluctuations in demand may make it difficult to identify marginal costs of particular passengers or freight. The demand for the transport of fruit and vegetables or grain may differ quite considerably from one year to another. When the railways carried large volumes of this traffic, they would keep sufficient capacity to meet it, though sometimes it might not have been used.

Two or Multi-Tier Tariff Pricing

Two or multi-tier pricing refers to the idea that in some cases, as for example with the Network South East card, there are two different charges to the passenger. The first is the cost of the card and the second is the amount paid for each journey. Ideally, transport operators would like to recover their fixed costs perhaps using such cards, whilst the other payment would cover the marginal or variable costs to the undertaking. The idea is not confined to passenger transport. There may be variations for freight transport (e.g. a lower charge if more items are sent). To some extent, the Post Office does this with bulk orders.

CHAPTER 7

Railways

Railways' Cost Structure and Reduction

Railways' fixed costs are high, as they include track, bridges and tunnels, track maintenance, passenger stations as well as locomotives and rolling stock. The railways in the UK and elsewhere have tried to reduce passenger costs often by closing down passenger stations and sometimes sparsely used railways completely. One of the criticisms is whether cost benefit analysis (CBA) should still be remained and applied for the assessment of the varied impacts of complex mega transport projects, as social, human life, environmental and built environment impacts can be overlooked (e.g. see a research paper conducted by Dr Robin Hickman and Dr Marco Dean (2018) – "Incomplete cost – incomplete benefit analysis in transport appraisal").

Railway Fares

There are a number of different railway operators in the UK and each of these has the power to set fares within prescribed maximum limits. Within England, fare increases are based on the retail price index (RPI) in the July of the preceding year. This applies to regulated fares which are mainly those for commuter services. However, for unregulated services there are far more options and some railway fares can be very low, especially if people choose to book far enough in advance. The current government has tried to ensure that a greater proportion of railway costs is borne by the passenger rather than by subsidies. Usually passengers in order to gain cheap fares have to specify the train they will be travelling on. This can lead to low load factors, since if the fares are very low, people may book on 3 or 4 trains as it will still be cheaper than the full railway fare.

Other problems can arise if connecting services are late and the passenger therefore cannot get on the service he or she desires. Sometimes, sympathetic staff are empowered to allow passengers some discretion, but in other cases a penalty fare of £20 or twice the full single fare can be imposed whichever is the greater.

Contactless payments can now be made for tickets which reduces the time taken to get through the ticket barriers. Ticket barriers have been used at many busy stations such as Victoria Station in London, but can pose problems for many people who have considerable volumes of luggage or are carrying pushchairs, etc. Online ticketing is frequently used by many passengers and reduces queues at booking offices or ticket machines.

Reducing Station Costs

Sometimes the cost of stations can be reduced by having unmanned passenger stations but with robust shelters, which keep the passengers out of the way of driving winds and rain. Increasingly passenger tickets are available online so that the costs of booking clerks have been reduced. Conductor guards have been used on many trains, which gives reassurance to passengers as well as reducing the number of passengers who sometimes deliberately defraud the operators by not paying the correct fares. There has been controversy about whether the railways need a second member of staff on the train for safety reasons (with regards to the debate, they also have several strikes on this).

Ideal Services

On both passenger and freight services, the ideal is to have well used fast and frequent services but these ideas conflict.

Conurbation Traffic

In the conurbations including London it is possible with both the Underground, London Overground and many suburban services to run a very frequent service. However, on some routes a frequent service would obtain low load factors so this would not be commercially viable.

Freight Costs and Revenue

A year after the Beeching report was published, a more positive approach to freight was published entitled "the Hoyle report 1963" that helped to set up the Freightliner network amongst other services. Currently, the UK's railway operators carry 11% of the UK inland surface freight. Coal has been the traditional market but this will fall because of climate change considerations. However, it is hoped that this will be offset by the change transporting biomass. Greater electrification of the railway system should also help reduce the railways' costs.

The Railways' Freight Operators

There are five main railway freight operators in the UK. Direct rail services are one of them and have a base in Carlisle. It is owned by the Nuclear Decommissioning Authority and is therefore currently a public-sector organisation.

Railways and Investment

The railways in the UK have often complained about underinvestment both before they were nationalised and also after privatisation. The 1950s modernisation programme had not considered properly the importance of electrification and the last steam engine was built in 1960. In contrast, countries such as Japan built the Tokaido line, which uses the train often referred to as the bullet train and subsequently the Japanese have built a large-scale infrastructure with modern punctual trains. Similarly, France with the TGV lines has developed a frequent long-distance passenger train service whilst still nationalised and Deutscher Bahn another nationalised industry has also delivered a frequent long-distance rail service.

Modern Rapid Transit Programmes

Hong Kong whilst still a British colony developed a rapid transit system from 1979 onwards which originally had round 3,300 passengers a train, 384 of whom were seated and 2,916 standing and with a frequent service gave a total of around 45,000 capacities in both directions. Whilst the average speed was only around 20 mph, accessibility to stations was very good, as no station was more than one and a half miles from the next station.

Underground Services

They have a number of advantages for cities. Unlike major roads, they do not isolate communities. They take up very little scarce land space, through it is possible to observe this with the new Elizabeth line in London and compare it with any road development in London or elsewhere. They usually give passengers covered access from the weather and so passengers are unlikely to be adversely affected by weather conditions. They reduce the effect of carbon emissions depending partly on how the electricity is generated. This is very important with the emphasis on climate change in most nations.

There is usually no conflict of line movements, since different Underground lines can cross each other at different levels as can be seen at King's Cross and St. Pancras where the Piccadilly line, Northern line, Victoria line, Metropolitan, Hammersmith and City and Circle lines all provide good access to the main railway network. Thameslink whilst not called an Underground service, is underground for the part of the journey which links King's Cross and St. Pancras with Blackfriars on the District and Circle lines. Blackfriars is an unusual station since it now straddles the River Thames. Underground trains are much quieter than other modes of transport. They are usually very safe although some people have used the 3rd rail system to commit suicide. The Moorgate crash in 1975 on what was then the Northern City branch of the Underground has never been satisfactorily explained. The King's Cross fire 1987 on the Northern line has been explained by the fact that at that time, escalators were made by wood and a match was dropped causing a fire. Following this, smoking in Underground is banned. There were also terrorists' attacks on the system in July 2005 on both the London Underground and the bus system. The major disadvantage of Underground systems is that they are very costly.

One of the problems however with the London Underground is that the mainly Victorian and early twentieth century infrastructure is having to cope with a massive expansion of traffic. The management of Transport for London (TfL) therefore wishes passengers to have very easy fast access to the different Underground and Overground services with tickets being bought either online particularly for season tickets and increasingly ticket machines using Oyster cards etc. to cater for the other customers.

Railway Staff and Training

In 2017, the UK railway industry employs nearly 200,000 people. In spite of unemployment in the UK being over 1,300,000 (which is still lower than for a considerable period) there are still problems with recruiting staff with sufficient skills.

The National Skills Academy for Railway Engineering was established in November 2010 to help reduce some of the problems the railway faces. Their "Routes into Rail" website highlights some of the skills which new recruits will need to have.

Apprenticeship Schemes

Network rail provides apprenticeships in track inspection as well as overhead line electrification, signalling and telecommunications. Women are still underrepresented in the senior engineering part of the railways system. The macho image of engineering does not help. It is important that careers officers in schools also make people aware of apprenticeship schemes.

Tunnelling and Underground Construction Academy

This Academy was set up in 2011 at Aldersbrook sidings near Ilford in Essex mainly to prepare for the construction of Crossrail. Crossrail has subsequently been renamed the Elizabeth line, but the Academy should also be helpful for other developments such as the extension of the Bakerloo line to Battersea.

Need to Study Ergonomics

Ergonomics is important in most of the transport sector both for passengers and drivers. It looks at the way in which people move their bodies and how they react. It is important that drivers can see signals but also can observe what is happening such as people or animals crossing the railway track, which could be dangerous for the drivers and passengers but could also lead to considerable delays on the railway system.

Other Vocational Examinations

The Chartered Institute of Logistics and Transport (CILT) runs a number of different courses which would be relevant for the railway industry. It is often important that people within the railway industry are aware of the other modes of transport, because they are sometimes complementary such as providing services to and from the railway network. They are sometimes competitive and all business organisations need to understand the strengths and weaknesses of their rivals. The Institute of Transport Administration (IoTA) also run relevant courses for the industry. Similarly, the Chartered Institute of Personnel and Development (CIPD) runs many courses in human resource management.

Catering for the Disabled

The Equality Act 2010 in the UK tries to ensure that people can carry out their jobs, etc. without discrimination. If people with disabilities cannot get to and from jobs then this is very unhelpful. Access therefore to and from railway services is essential. Many railway trains can now accommodate mobility scooters which is helpful. Many modern trains also have disabled toilets, although this adds considerably to the cost of new trains. The same applies to new stations. Sometimes, however, the height of ticket machines at railway stations makes it difficult for disabled people to be able to use them.

Railfuture (the Growing Railway)

Railfuture (an independent pressure group) published a book in 2017 entitled "Britain's growing railway". This lists all the stations, which have been reopened in Britain and the original passenger traffic at the time of reopening and the current traffic. Some of the stations which have been reopened have obtained massive increases over the forecast traffic.

Government-Control of the Railways

The Department for Transport (DfT) has overall control of the railways in Britain. Modern Railways a well-respected magazine which is independent of the industry commented in an article in November 2017 that ironically there is now more control of the railways by the UK government than when it was run as British Rail as a nationalised industry. The current Secretary of State for Transport is "The Rt Hon Chris Grayling MP", who was appointed on the 14th July 2016. He has been heavily criticised for backtracking on electrification of the railway system.

The Rail Franchising System in Great Britain

The railway franchising system in Great Britain means that passenger railway operations are run through a system of franchises to companies. This system only applies to Great Britain since in Northern Ireland there is a state-owned company Northern Ireland Railways. The system was created under the Railways act 1993, and prior to this the railway system had been owned by British Rail (BR). The Treasury followed a plan which had been advocated by the Adam Smith Institute which contracted passenger services to seven-year franchises. The track and signalling was run by Railtrack but the rolling stock was sold to rolling stock operating companies (ROSCOs). In turn the rolling stock operating companies leased rolling stock to train operators. The franchises were put out to competitive tender and this was originally overseen by the Office of Passenger Rail Franchising (OPRAF). Most franchises are run in terms of a geographical area for example Southeastern covers the areas of Kent and East Sussex as well as part of London.

There is also a West Coast mainline service, which covers the route from London to Glasgow. This ran into controversy, because it was massively over budget and also considerably ran over the time allocated for the upgrade. Originally the DfT had in 2015 offered the InterCity West Coast franchise to FirstGroup but in October 2012, Virgin was offered a franchise for a short period. The logic of privatisation was partly that the private sector would bear the risk and any losses. However, there are clauses in newer franchises which offer sometimes compensation for lower-than-expected revenue and also clawback any excess profits.

Many critics think that it is strange that nationalised industries from France (SNCF) and Germany Deutsche Bahn and the Netherlands (Spoorwegen) as well as from China have been allowed to run train services in this country, whereas the publicly owned sector including East Coast had operated from 2009 after National Express East Coast had defaulted on its contract. The Chinese firm MTR is helping operate UK's South West trains franchise.

Similarly, Connex South Eastern a French operated company was stripped of its franchise in 2003 and instead a publicly owned Southeastern Trains took over. Whilst the majority of railways in the UK are run as franchises there are also some others which are run as concessions. This includes TfL's overground which has widely been perceived to be successful. The percentage increases in demand have been very high for some stations.

There is also the Docklands Light Railway (DLR) as well as Manchester Metrolink. The difference between a concession and a franchise is that the concession holders are paid a fee to run services where the details of service frequency, etc. are usually specified by the authority which awards the concessions. Prior to 2011, the franchising system had what was often described as a "cap and collar system" whereby the franchisors and the government shared the risk about future demand.

Labour Party Proposals

The Labour Party leader Jeremy Corbyn has stated that a future Labour government would take over franchises and put them into the public sector, as they came up from renewal rather than a swift wholesale renationalisation.

Christian Wolmar's Comments

The book "Broken Rails: How Privatisation Wrecked Britain's Railways" by Christian Wolmar published in 2015 describes how privatisation wrecked the British railways system, he suggests that fragmentation of the railways has been a complete disaster. Not everyone will agree with this view, however, the idea that competition will automatically bring down the subsidies paid to the rail sector has been untrue. Allowing for inflation the subsidy paid to British railways as a whole has increased. The complexity of the franchising system has not helped.

CHAPTER 8

Reasons for Government Intervention

The term 'government' is used here in the broadest sense to include the EU Central and Local government. Intervention occurs for a variety of reasons including safety, control of monopoly and increasingly the protection of the environment. The government may also intervene to prevent wasteful competition, for strategic reasons, for reasons of prestige or in order to improve balance of payments. It may also intervene because of the divergence of social and private costs and for long or short run manipulation of the economy. In both developing and other countries, investment in transport infrastructure has been assumed to lead to economic growth.

Intervention for Safety Reasons

Methods of intervening in industry for safety reasons vary considerably. It may include quantity licensing on the assumption that if there is free entry into the market profits may be reduced to the bare minimum and therefore safety standards might fall because firms skimp on maintenance. This was one of the reasons for intervention by the UK government in both road freight and road passenger transport. There is little evidence to support these assertions. The government may use detailed regulations such as the construction and use regulations, which apply to both road freight and road passenger vehicles.

There are often similar regulations for other modes of transport. The government may have a licensing system for managers, as in the EU countries. The certificate of professional competence applies to both the road freight and road passenger sectors. In most modes of transport, there are also controls over crew, such as air pilot licensing, public service vehicle and driver licenses and heavy goods vehicles licenses. There may also be controls over "The Way" whether the government provides this directly as with roads or indirectly through the nationalised industries such as the railways. One of the criticisms of government control over safety is the government does not control safety in a systematic way. In particular, regulations are more stringent on some modes rather than others. This partly arises because the public are much more affected by the occasional spectacular air or rail crash, even though generally safety is much better per passenger kilometre than for road transport.

There is often little analysis of the cost of the variety of measures; however, it is relatively easy to find out how much hours and regulations have cost the road freight industry and in the UK, the Freight Transport Association has published figures about this. It is more difficult, however, to find any evidence about the reductions in hours affecting safety. In the UK, the reduction in hours took place between 1978 and 1981. Sometimes there is a political unwillingness to accept certain safety standards. There is considerable evidence about drinking and driving. The Blennerhassett report to the Home Office in 1975 wanted more stringent measures, but these have not been accepted partly on the liberty of the individual grounds. This is rather ironic as drinking certainly impairs other peoples' liberty and also because such controls would be far cheaper than investing in new infrastructure. The majority of benefits of new roads are improving safety.

Intervention to Prevent Monopoly Abuse

The government has often controlled fares and charges mainly because of monopoly fears. This applied to the UK railways after 1844. The controls have sometimes been imposed a long while after they were necessary. British Railways had fares and charge controls up until the 1962 Transport Act, though by that time it faced a considerable degree of competition from other modes of transport and was rarely a monopoly freight supplier. One of the problems of imposing price controls is that because transport is not homogenous, it is difficult to set maximum rates, which might not prevent useful services being operated. The government has often made provision for compulsory purchase of land whether for railway track, canals, airports or ports. This is because of the importance of transport which may outweigh individual property rights. A current example of this is the dispute about increased airport capacity at either Gatwick or Heathrow Airport. Disputes have also arisen with the proposal for High Speed 2 (HS2) from London to Edinburgh and Glasgow.

Government and Protection of the Environment

The government may also intervene to protect the environment. It can do this through fiscal policy encouraging use of battery or hybrid operated vehicles as in the case of the UK where since the early 1980s no vehicle excise duty has been paid by battery operated vehicles. Alternatively, legislation can be employed as with construction and use of regulations governing the legal noise limits of lorries and other vehicles. At the international level, there have been agreements about the amount of noise which modern aircraft can make. The price system has sometimes been used for example the British Airports Authority charged in 1985, 25 per cent more for aircraft which did not meet the required noise limits.

Government and Strategic Reasons

Governments also intervene for strategic reasons. Roman highways were often built for this purpose and the term highway itself indicates that the roads were often built above the level of the existing countryside in order to avoid ambushes. Some of the support for research and development given to aircraft manufacturers has been justified partially on strategic grounds. This poses a difficult problem for the economist in trying to judge whether such money has been well spent. It is sometimes alleged that the build-up of the Russian Merchant Fleet including passenger liners during the Cold War had been partially justified for strategic reasons. The Tanzam Railway linking Tanzania and Zambia in 1970 to 1975 was partially built to avoid originally hostile regimes such as South Africa during the apartheid years so that Zambia was not reliant on white ruled such regimes for its transport.

Governments and Prestige

Concorde was built partly for this reason and partly because of beliefs in the importance of technology in the 1960s. No published cost benefit analysis was made about the likely effects for demand for such an aircraft. It is often alleged that many Third World countries have their own airlines for prestige, rather than for other reasons. There may be other reasons such as balance of payments.

Governments and Infant Industries

The logic of subsidising infant industries is that without such protection they might not achieve later economies of scale where they could become self-sufficient. This could apply to the aviation or shipping industry or vehicle manufacturing where more efficient vehicles need time to become established.

Economists have often been scathing about the so-called Peter Pan nature (i.e. that the industry never grows up), but again it is difficult to find evidence to see whether the industries have in fact grown up and whether subsidies or protection no longer apply.

Governments and Balance of Payments

The increase in oil prices after 1973 meant that many countries should have taken the conservation of fuel more seriously though, too few companies and countries have had consistent transport policies to allow for this. Transport economists would have expected the increase in oil prices to lead to a more systematic valuation of how to reduce the amount of oil used. This could have included greater efficiency of existing systems such as managers using tachograph data and computer scheduling. The government could also help through providing information so that return journeys were not made empty. There have been suggestions that the UK government should inaugurate a system of brokers whereby knowledge of return loads was made more freely available to road hauliers and similarly that potential car passengers and drivers could be put more readily in touch with each other. The alternative would be to use non-fossil fuels. There could be greater emphasis on battery or hybrid operated vehicles. Similarly, electrification of the railways which would not then be reliant on one source of fuel, would be helpful. It is possible to find comparative electrification figures for railways. The UK has lagged behind many other European countries. Reducing fossil fuels may help many countries with their balance of payments problems. Many countries have subsidised domestic shipbuilding or aircraft industries or encouraged transport in order to boost tourism. However, greater transport provision may have an adverse effect since it may also lead to more imports and home tourists visiting other countries. Some governments have also intervened for balance of payments reasons by having systems, which make exports cheaper and imports more expensive through manipulation of freight charges.

Government and Economic Development

In the early stages of industrial or agricultural development transport provision may be justified on these grounds. Many West African countries had railways going to the interior in order to provide raw materials for the colonial powers. In order to justify such development economists would wish to have assurance that the development would not have taken place in the absence of transport and that the transport was provided cost effectively.

Employment

With the increase in unemployment in many Western European countries including the UK in times of recession, it has been suggested by both the Federation of British Industry and by the Trade Union Congress that an increase in transport infrastructure could alleviate some unemployment problems. Even the Channel Link, which during the 1960s was often claimed to be labour saving, has been promoted by the rival consortia as leading to a large number of construction jobs. This is one of the reasons why the late Margaret Thatcher, Conservative Prime Minister and the late President Mitterrand agreed to the construction in 1985. The assumption also was that there would be a boost to manufacturing in the depressed areas of the UK and in France. Opposition to such development, however, arose from Dover and other ports because of perceived long-term job losses there. Dover has, however, partly reinvented itself as a cruise ship terminal.

Government and Long Term Fuel and Other Problems

The government might look at the electrification of the railways, which would not then be reliant on one source of fuel, apart from the effect of oil price rises and the subsequent effect on balance of payments. Governments may also intervene for balance of payments reasons to subsidise domestic shipbuilding or aircraft industries or to encourage transport to boost tourism. However, greater transport provision may have an adverse effect since it may also lead to more imports and home tourists visiting other countries. Some governments have also intervened for balance of payments reasons by trying to have systems which make exports cheaper and imports more expensive through manipulation of freight charges.

CHAPTER 9

Road Passenger Transport – Buses, Coaches, Taxis and Cars

Reducing Costs

All transport organisations wish to reduce staff costs if this can be done cost effectively without compromising on their objectives.

Why Did One-Person Operation Come in on the Buses?

From the 1970s onwards, there was a trend to one-person operation. This was partly because wages were a high proportion of total cost, in 1973 just before the Oil Petroleum Export Countries (OPEC) price rises they were estimated to be about 70% of total costs.

Staff Shortages

Often there were staff shortages. London Transport complained that it was about 6,000 staff short when it employed about 60,000 staff. Other municipal operators claimed a much bigger percentage shortage (i.e. Southend-on Sea claimed that it was about 30% short). Several different methods could have been used to overcome the shortage. One of the methods of doing this was to have one person-operated buses. Higher wages, could then be paid with the reduction from 2 to 1 staff. Shift work, which was necessary to cover the peaks, is unpopular. There was also the problem of staff who do not own or have access to cars getting to and from work.

One of the problems for many operators was that if they did succeed in raising wages and gaining more staff, this might be at the expense of staff shortages elsewhere. There is here the fallacy of composition argument that whilst one undertaking might be able to attract more staff because of this, it does not follow that if all undertakings did this, that the staff supply would be much greater. A rise in wages would have a much greater effect on the elasticity of supply for one operator than for operators as a whole. In the 1970s, there was often hostility towards women drivers, both from the management and from trade unions. Another possibility was to allow part time work, but this again would have been unpopular.

Changes in the Design of Buses

The advent of a one-person operator operation also meant that much greater emphasis was placed on the design of buses so that often instead of the traditional entrance at the rear there was an entrance by the driver who could then check the tickets in some way and also a central exit. It also gave a greater incentive to having tickets issued off the bus and currently buses operated by TfL have all tickets issued off the bus.

Productivity of Buses

One of the problems for bus operators is that passengers generally prefer a door-to-door service or an approximation to this. This would require many bus stops. On the other hand, the greater the distance between stops the greater productivity of buses. Bus operators could help increase productivity in different ways. Some are internal and some are external. Internal solutions include better timetabling, using express buses in one direction so that buses have a better chance of running to time and also more likelihood of being able to use the buses for more journeys.

Bus Lanes

Bus lanes have become more common, although the advantages are usually only justified if the number of buses using them is large since they reduce the available space for cars. If there are enough buses, the scheme is more likely to be self-enforcing whereas if there are relatively few buses, motorists are more tempted to go into the bus lanes. It is noted that anyone can use the bus lane outside the hours of the operation from 7.00pm to 7.00am at night. There are also in some areas contra lanes where buses can go against the general flow of traffic. One of the problems of the bus lanes has been that of enforcement.

London Transport in the 1970s and its successor TfL currently have often complained that regulations have not been enforced sufficiently.

Another method would be to reduce parking or eliminate it in areas where buses are a major form of transport. This is sometimes unpopular with some local residents but makes much better use of the way. Red routes in the London area restrict parking.

Bus Bays

Sometimes there are parts of the road where the buses will be allowed to park, which should not impede other traffic. However, bus operators have commented that apart from being comparatively expensive, buses have then found difficulties in getting out into the main road.

Priorities at Junctions and Elsewhere

Delays can occur at traffic lights and other junctions. There have been some experiments with this, and they show that the delays can be approximately halved thus increasing reliability of buses.

Bus Roads

Runcorn in North West England has many bus only roads and the speeds have typically improved to around 22 mph (35 kph).

Size of Buses

If bus tickets can be issued off the bus, then this presents fewer restrictions on bus size. Otherwise, the size is limited by the time loading the bus.

Double-Decker and Single-Decker Buses

In the UK, Double-decker buses have been much more common that in most other countries. Double-deckers have not been used in rural areas or where low bridges prevent them. Double-deckers are more common partly because on most roads in the UK the headroom is around 5 metres, but is lower in many parts of Europe and the USA.

Double-deckers carry more passengers within the same length of road. Therefore, they have been useful in major cities, such as London and Birmingham where traffic flows are heavy. The disadvantage is that not all the space is available for seats and there may be problems for older people or those in wheelchairs. Modern buses are generally better and easy access buses, which are at the same height as the pavement, make getting on and off the bus much easier for passengers such as older people, wheelchair users and parents with pushchairs. The present government has tried to ensure that disabled people can travel on buses so that there is now more accommodation for disabled to get on buses with wheel chairs if necessary.

Articulated buses have increasingly been used, and can carry a maximum of about 150 people. They are more widely used on the continent although "bendy buses", as they are often called, have gradually come into the UK. The "bendy buses" withdrawn from London in 2011 because of the safety and fire evasion issues. Where passengers require a fast-reliable service, it is possible to have buses with relatively few people sitting, but more people standing.

Guided Bus Ways

There have also been some experiments with guided bus ways. The advantage of guided bus ways is that the bus, because it runs absolutely next to the pavement, takes up less width than in a conventional road. Cambridge County Council has expressed an interest in a much longer one than the Ipswich one.

Decline in People Using Bus Services

One of the hopes of the 1980 and 1985 Transport Acts expressed by the late Professor John Hibbs was that small new organisations could enter the industry. This has not happened generally in the UK. Information about the size of operators can be found in Transport Trends. This may be because, as the number of cars rises, congestion also increases so that the operator is faced with both higher costs since the productivity of the buses worsens with congestion and also whereas car drivers may be able to take so-called rat routes this is not generally an option open to the operators. The fares are likely to go up, since there are fewer people to share the fixed costs of the mode.

The government announced in 2001 that there would be a Bus Urban Challenge Fund, which would be open for county councils, unitary authorities and county councils as well as Transport for London to bid for and the initial amount allocated for this fund was £46 million. The fund could be used, to allow for better quality buses to improve accessibility. This would be helpful to the disabled as well as parents struggling to get pushchairs on and off buses. There would also be much clearer electronic boards on buses that show more clearly the bus destination.

Information and Buses

A few bus-stops now have an oral system, which will say when the next bus services will come. It is hoped that this will be of use to blind or visually impaired people who if travelling by themselves will not be in position to know about possible waiting times or delays. In many UK areas, passengers can text to find out about bus services.

Limited Stopping Service

These are used by many operators since a) for longer journeys the greater accessibility times are often going to be outweighed by the faster journey time, and b) the operators can get better utilisation of their fleet especially if, as happens in many continental countries, the buses are not usually delayed by the issue of tickets.

Some bus operators run buses with a limited number of stops. This can be very helpful in the morning rush hour and to some extent in the evenings. Sometimes the buses will not put down before a certain area. This is an advantage to the longer distance passengers, since otherwise they find themselves squeezed off the buses by short distance travellers. Limited stop services may be relatively infrequent because of shorter distance passengers. It has the advantage to the bus operator that sometimes they may be able to avoid some congested parts of town, by operating on a different route. In Birmingham, currently some limited stop bus services can use the urban motorway for part of the journey. From the passengers' viewpoint, it will usually give a quicker service since acceleration and deceleration takes time.

One problem however, is that it may limit the overall frequency of the service, so the operator has to weigh up the benefits of the faster service against this disadvantage.

Coach Services

The coach service can be subdivided into three segments. One of these is the express journeys, which not only serve the major towns but also serve places which were left off the rail map after the Beeching cuts in the 1960s. The National Express Service which was part of the then National Bus Company offered very cheap fares on some of the longer distance services such as that from Birmingham to London in the early 1980s. Because these fares were available at any time of the day, this was convenient for those people who might have fairly early morning appointments. Additionally, they sometimes stopped at an outer London Underground station, such as Hendon so that the disadvantage of going to London and the subsequent London congestion could be avoided.

A second service is the tour operators, which may be one-day tours. Many one-day tours served the seaside. Nowadays, however, one-day tours are more commonly longer. The shuttle service to and from France is helpful to people who do not fancy the bad weather which can affect sea crossings. Sometimes the coach operator will operate these services on their own behalf and sometimes they will be run by tour operators who then subcontract their work to the coach operators. The advantage of the international tour operator from the passenger's viewpoint is that they will be helpful to many people who do not have the necessary skill or possibly the desire to look up the variety of destinations which the coach operates. In these or the more mountainous areas, there are not always easy ways of reaching to these areas otherwise and the problems of looking up timetables in other languages can be a daunting one. These operators additionally have the advantage of bulk purchase of hotel rooms. The advantage of the coach here is that people can keep their luggage with them which can be a major asset especially given that the railways no longer offer the comprehensive "passenger luggage in advance" facility which was very helpful to holiday makers but also to many others visiting friends and families. A great many services go to Victoria coach station but the contrast between Victoria coach station and the major airports such as Heathrow could hardly be greater. Victoria coach station is very cramped and offers minimal facilities. Victoria coach station whilst not far from the Underground network has no direct line to it or direct links to the rail network. Relatively few people will wish to stay in the Victoria area because of this inconvenience.

Immediately after the 1980 and 1985 Transport Acts many commentators hoped that the companies might be able to compete through lower fares for many journeys for both long distance journeys such as those from Birmingham to London as well as from the commuter routes where the advent of cheaper fares might have been an attraction. However, road congestion in the London area means that relatively few people in the London areas go to work in central London by coach. The coach service frequency does not usually compare favourably with the train services. Whilst fares are considerably cheaper than the rail service, if people have to work late or for other reasons wish to stay in town, they often find that there is a minimal service in the evenings. Most long-distance coach services do have the advantages of providing toilets and often provide a drinks service and occasionally a light meal service. With a few exceptions, UK coach services are not well publicised. This is in contrast to the Greyhound services in the USA, even though the coach operators in the UK carried far more passengers in the UK than the Greyhound services do in the USA.

Private hire coach services are important for organisations including educational authorities. This may include providing the service to and from schools particularly secondary schools and occasionally colleges where the standard bus or occasionally train service is inadequate. This is often done by smaller organisations since they do not have to be concerned about marketing. This is often a weak point of many smaller and some larger organisations.

Trolleybuses

These no longer run in the UK since the last one ran in Bradford in 1972. They are like ordinary buses but with two spring-loaded poles attached to them, which then reach to electrified overhead wires. They therefore have some flexibility compared with trams. Worldwide they first operated in 1882 but did not come into ordinary working until 1901 and in the UK from 1911. They had the advantage of being a quiet form of transport since unlike trams, they have rubber tyres and also unlike trams they are very suitable for hilly areas. In London, trolley buses went up steeper hills including Highgate Hill, which the trams could not reach. In 2012, there were about 300 trolleybus systems worldwide.

Dial a Bus Services

Dial a bus services are for disabled and old people with reduced mobility problems. This can be used where the bus will follow a fixed route generally, but will divert if the person before has either phoned through to request the service or sometimes for the relatively few people who do not have a phone they may pre-book possibly at a shop or Post Office etc. The advantage of this from the passenger's viewpoint is that it gives a more direct service, whilst from the bus operator's viewpoint it means that they do not have to divert sometimes from the main service run in order to have the possibility of picking up passengers. If people know of the service therefore, it has the advantage to the operator that it reduces the amount of fuel that they use and will usually give a slightly quicker service.

Problems of Reductions of Services

Sometimes managers have not understood costing. This was a criticism of the Beeching Report but the same has often applied in other industries including the bus industry. By looking at average costs rather than avoidable costs, operators would look only at the average cost and would therefore try to reduce costs by cutting off the furthest points of the transport network. This particularly applied in what was sometimes called the Christmas tree routes (i.e. where the main route was along a main road, with others branching from this to outline villages). Bus managers therefore, found that often that they would cut out these outlying villages without realising that some of the bus services in effect were acting as feeders to the other routes. This made the average cost higher and so the high prices became a vicious circle.

It was often assumed that the peak was profitable whereas the off peak was unprofitable. This was untrue, but often meant that the frequency of service in the off-peak hours was sub optimal and therefore meant that vehicles were not operated where often bus operators could have offered a more attractive service and retained passengers utilising drivers better during their shift. Choosing a transport mode is often a habit. Therefore, if the transport network cannot offer a comprehensive network, in the eyes of the passenger, this may lead to a change in habits. Once potential passengers realise this and as cars have become progressively cheaper to run compared with public transport, notwithstanding fuel protests, the service becomes less attractive. Currently, operators are in a ratchet situation (i.e. that once people have made the decision to take a car then the car owners are less likely to use public transport).

Similarities of Bus and Rail Industries in Looking at Costs

Often using the average cost data, it was not realised that revenue in the reverse direction, given the joint cost argument, was often worthwhile having. If the bus or train is running and is likely have low load factors, then any additional revenue is worthwhile having. The marginal cost of any additional passengers is usually negligible. Therefore, even within the peak hour traffic, particularly going to London and other conurbations, it is often worthwhile offering cheap fares in the opposite direction. This is particularly true in London where some commuting routes are from attractive places, such as seaside resorts including Southend-on-Sea and Brighton and therefore cheaper fares in these periods and the high frequency of services might attract more customers.

Open Top Buses

In some towns, open top buses have been used. Traditionally they have been used at seaside resorts such as Southend-on-Sea and Bournemouth for a number of different reasons. In Southend-on-Sea, it was partly because a low bridge at one point would have prevented conventional double-decker buses going under it, but the capacity of a double-decker bus was needed for the number of tourists if costs were to be kept down. The open top buses also had the advantages of attracting passengers, perhaps because the British allegedly like the open air. Sometimes the buses could have been older ones, but people in the UK are interested in either novelty or new. Open top buses are novel. The fixed costs of the system are therefore negligible since often the buses have been fully depreciated so that there is no fixed cost for this. The fuel cost may be slightly higher than with newer vehicles and the fact that the vehicles are used much less than many other buses is irrelevant given the low fixed costs.

In towns, such as York and London, the open top buses are popular partly because they give good views of historic buildings, which could not be obtained from any other mode of transport. Sometimes this is supplemented by commentaries, which are available in a number of different languages. The frequency of service is usually such that by operating a limited number of buses they can offer a reasonably frequent service by issuing all day tickets. Whilst the demand is somewhat unpredictable because of the British weather, they do offer the opportunity of providing very good information at least about the places they are visiting. Additionally, where car parking is both scarce and expensive it may provide a welcome alternative to the car, which is of benefit to the passengers and to the community as a whole.

Taxis

Drink drive legislation has become tighter in the UK and it seems likely, it will be tighter still, and then people have a number of choices. They could forego going to the pubs in the first place, but this is somewhat unlikely. Alternatively, one person will not have a drink during the evening and will act as chauffeur (i.e. the designated driver) to the other people car users. The other possibility is that all the people can drink if they wish and then have taxis in both directions.

Economists might assume if the then Norwich Union type of pricing had become more common, then young users might use more taxis in the small hours. The Norwich union had imposed very high premiums for young drivers in the small hours, because of the very high risks involved. The taxi fares could be lower than the perceived cost of motoring. Given the high accident rate for young people this would be desirable.

Apart from the conventional evenings, taxis are often used at special times of the year such as New Year's Day or Christmas Day when conventional public transport has often closed down. Currently London Underground is now offering tube services throughout the night on Fridays and Saturdays on the Victoria, Jubilee, and most of the Central, Northern and Piccadilly lines.

Taxis are a neglected part of the public transport sector, although they can be important for different reasons. The first is that for holidaymakers carrying a considerable amount of luggage the taxi to and from their local station as well as from the destination may enable them to make journeys, which would otherwise be difficult. There is also a second need for the service for example in some areas business executives, in particular may want to have a faster service using possibly the rail or air services as part of their total journey and the taxi gives the advantage of both privacy and also speed. Sometimes it may actually be quicker than using ones' own car, especially where there are problems of parking. The conditions for taxis vary tremendously between different local authorities. In London, the black cab taxi drivers have to take the test, which is often known locally as "The Knowledge" (i.e. they are expected to know the different streets of London and how to get there in a sensible way). This is apart from the usual safety requirements for taxis. In other local areas, there have sometimes been complaints that there are insufficient taxis because there has been a quantity licensing system.

In New York, there are a limited number of licences available, so therefore potential taxi drivers will have to pay considerable sums in order to obtain one.

Basis of Charging by Taxis

Where local authorities specify fares, there is usually a minimum amount which is payable irrespective of distance. For example, this could be 1 thousand yards (approximately up to 1 km) and then so much per mile beyond this.

The disadvantage of this from the taxi operators' viewpoint is that short distance fares even whilst they may seem excessive to passengers will often not compensate for the amount of time they are waiting around in taxi ranks. Often longer distance journeys fares are subject to negotiation by the taxi firms.

Structure of the Taxi Industry

The taxi operators vary from at the smallest end of the scale people who own their own taxis to large fleets. The advantage of the owner-driver from the owners' viewpoint is that they do not waste time, in many cases getting to and from a set point, but can operate from home and if they are waiting can utilise this time more effectively than if they had to wait at a typical office. The disadvantage is that it is often difficult for them to get good utilisation of their taxis since sometimes they may be busy and could really do with several taxis, whilst at other times they may be slack. The theory of large numbers will illustrate why they will get poorer utilisation than a larger firm. There are also fleet taxi operators. Sometimes the drivers own their own vehicles, but will share their revenue with the company, which operates the fleet using an office with a telephone connection. The advantage of this from the taxi drivers' viewpoint is that in effect it takes account of the marketing which is otherwise difficult for a small firm, whilst still giving them some flexibility (i.e. they can choose not to work certain days if they do not want to do so). On the other hand, by having to share the revenue, they do not get as much as if they operated all the journeys for themselves. In other cases, the taxi fleet is actually owned by the operator. The operator in this case has greater control over the services and can usually get a fairly substantial discount by buying several taxis at one time which can be well below the standard list price. Larger firms will get better utilisation from the taxis since clearly if the control system is any good, they can try to ensure that a taxi driver going to a certain district will be able to pick up people from that district. In the UK and in some other countries uber taxis have entered the market. After some legal battles it has been decided that they are classified as taxi operators and their drivers are therefore classified as employees rather than self-employed which has implications for their tax and social security.

Maxi Taxis

There have been suggestions from time to time, that taxis should be shared in some way (e.g. Uber taxis (pool: shared ride, lowest cost) – people can share their ride in exchange for reduced cost). This might reduce the fares for passengers whilst giving more revenue to the operators and could reduce waiting time at busy periods. On the other hand, the journey time might be slightly slower for the people at the furthest ends of the journey. There have been some examples of this in London between railway termini.

CHAPTER 10

The Way

Importance of Flat Land for Railways

Historically the railways wanted flat land, since even an incline of about 1 in 70 was severe for steam trains. The coefficient of friction of steel on steel is very low so that high speeds are possible with relatively little power compared with the road system. If standard gauge 4 ft. 8 ½" is used, then there will not normally be scope for tight curves and so ideally a straight line is required. This does mean that if there is an obstacle for whatever reason long curves are usually necessary. If flat land was not available, then there could be viaducts as with the famous Ribblehead viaduct on the Settle-Carlisle line or tunnelling, but either of these will add considerably to costs.

Standard Gauge

Originally most railways were built for the conveyance of coal and other minerals for short distances in the UK. Railway operators could therefore choose what gauge (i.e. the distance between the railway tracks that they wanted), however, as distances became longer, it was obvious that there needed to be connections between the different railways and most long-distance track used what is now called standard gauge which is 4 feet 8½" between the tracks. However, Isambard Kingdom Brunel used a different gauge for what became the Great Western Railway and this was 7 feet. Eventually in 1892 this was converted to standard gauge. Most countries in Europe also have standard gauge which makes it easier for trains to go from one country to another.

Different Types of Track Layout in the UK

It is possible to have a single track as occurs on some branch lines, but this reduces the capacity of the railway line quite considerably. Overtaking is impossible. Additionally, if another train wishes to pass in the other direction then there must be a passing loop. Double track, which is the most common system in the UK allowed for two ways running and more than doubled the capacity. On a few main lines, there were four tracks with the two outer usually being reserved for slower trains with the inner two for the faster trains.

When engineering work is required, it may be necessary to close one track (double track routes) which reduces the capacity for the duration or, if major work is needed, then complete closure of the route may be needed. This usually means bus replacement services to ferry people around the closed section or using a diversionary route. This option has become more complicated in the UK, since privatisation, as operators do not always have access rights over the alternative route.

The railways have sometimes used a narrow gauge. This often had a track gauge of about 2 ft. to 2 ft. 6″. This had the advantage that it could cater for tighter curves which meant it could follow the contours rather than having expensive tunnelling or viaducts. It also made the cost of track laying much cheaper. Narrow gauge was also sometimes used on industrial railways for the same reason. However, by their nature, maximum speeds were lower and transhipment from narrow gauge to standard expensive. Narrow gauge is not now found on the main lines in the UK, but there are 100 heritage railways such as the Ffestiniog Railway in the UK.

New Zealand is an example where the standard gauge is 3 ft. 6″ creating problems to upgrade the routes for higher speeds if the system requires motive power or carriages of rolling stock from other countries.

Railway Signalling

The cost of the track is not, however, the only cost and signalling can be expensive. Traditionally the railways used semaphore signalling (controlled from a staffed signal box) with horizontal meaning stop and diagonal meaning go. The advantages of the semaphore signalling were that it was easy to understand. This meant that distances between boxes were at irregular intervals. Track capacity was therefore limited to the longest distance between boxes on a particular route. As safety demanded that only one train was in a section (usually between two boxes) at a time, (known as block posts) capacity could only be increased to a limited extent if additional signals were provided in between boxes or extra boxes were provided. The railways have had a choice between having either more block sections, which would have added to the cost since more signals would have been made, or fewer block sections which means that hold-ups are more likely to occur. As signal boxes became larger they coped with a great variety of signals and often more signals can be put in with colour light signalling with only the cost of signals to consider. This was not the case with semaphore signalling. There are also often signals near level crossings.

To solve these problems, in the UK, there are now either 2, 3 or 4 aspect colour light signals controlled from a Power Box (to distinguish them from the older style). This is very similar to the road traffic light systems widely used with green meaning go, amber or double amber meaning caution and red meaning stop. On the railway, the choice of 2, 3 or 4 aspects is determined by the number of trains likely to use the line and importantly the maximum line speed permitted; the higher the speed the longer it takes to stop. Thus, double yellow (advance warning), yellow (caution), and red (stop) gives plenty of time for the train to be brought to a safe stop without having to keep line speed low to help ensure signals are not passed at danger. Power Boxes are more flexible, cover a much bigger area (there are only two power boxes controlling the line between Liverpool Street and Cambridge about 56 miles), signals can be placed at regular intervals and staff in the boxes have a much better overview of all trains in their area, enabling regulating the order of trains to be more reliable, such as keeping a stopping train behind an express. It has also enabled a large number of mechanical boxes (semaphore signalling) to be closed. The Hertford East branch was taken over by Liverpool Street power box enabling four local boxes to be closed. Modern technology also enables level crossings to be automated with remote monitoring from the newer boxes. This has made savings where a local box could be closed.

There is an even newer system known as "moving block". A train is protected at each end by a remote electronic "envelope" which moves with the train. Each train is protected this way and no two envelopes can overlap. This is enhanced by cab signalling to warn drivers to slow down if there is another train in front. It was intended to be included in the West Coast route modernisation, but it would have meant that only specially equipped trains or locomotives would be able to use the route and not implemented.

The railways have had problems with inter-sections since they often cause bottlenecks and one of the features of the railway plan announced in July 2007 was that more money was to be available to remove these bottlenecks.

Sometimes flyovers could be used as with the Eurostar service when it came into operation in 1994. The Eurostar could leave the main route via Bromley South into Victoria and travel via a new connection and flyover into Waterloo International station without having to cross the domestic traffic. Until the introduction of Eurostar services into St. Pancras in November 2007 the Eurostar had to compete with domestic services from Swanley almost into Victoria.

There are a number of alternatives, having more tracks would avoid the conflicts and flyovers to avoid this completely.

Railway Stations

On conventional double track, which is fairly widespread throughout the UK, the railways have a choice. Sometimes, they have island platforms which make it easy for platform staff to be able to supervise trains, since they do not have to cross from one platform to another. On the other hand, this means that passengers cannot get to the platforms without the use of stairs, which is not necessarily convenient to passengers. This problem has been recognised and lifts have become much more widespread at large and busier stations, but rarely at smaller stations. Costs of a footbridge and/or lift, even to cross a double track, may cost more than £1 million at current prices. The use of two separate platforms means that trains do not have to slow down even slightly when passing through the station nonstop and people can often transfer from street to platform without the need to use bridges or subways. In Germany, the majority of stations have at least the equivalent of a small conveyor belt going alongside the passengers so that luggage or pushchairs are more quickly conveyed. Historically, the railways had porters which added considerably to the railways' costs and therefore this practice was generally abolished. However, in some stations there are now mobility assistance service (e.g. Euston railway station in London; Manchester Piccadilly railway station, etc.) which will transport disabled people. The use of luggage trolleys has also helped some people.

Stations themselves are costly, although in Canada and in more remote areas, platforms are not used and passengers merely signal their intention to stop the train.

The railways have suffered continuously from dishonesty of passengers and often ticket barriers have been introduced which check tickets.

Unfortunately, however, these have often been unmanned which means that passengers who deliberately wish to be dishonest go to stations which do not have barriers. There is a correlation between dishonesty of passengers and other bad behaviour so that if the barriers can be implemented and manned, then other problems are reduced. In the UK, perceptions of risk have been magnified by the media. They have added to the disinclination particularly of women to use the stations, especially when they are likely to be alone. Conversely, in Switzerland, nearly all stations are staffed whatever their size and this has the advantage of overcoming the problems on the UK network.

Electrification

Electrification adds considerably to track costs. The Southern railway in the 1930s was mainly electrified using the third rail system, which has been widespread on the areas serving Kent, Sussex, Berkshire, Surrey, and Hampshire, but has not usually been used elsewhere. The third rail system has the advantage that it is often unnecessary to adapt the tunnels or bridges which is not true with the overhead pantograph system. The pantograph system which uses wires rather than an additional rail is generally cheaper, apart from where tunnelling or bridges have to be modified. It uses higher voltages and means that the power of the trains can be great. This is particularly important as with the domestic services in Kent, where it has been difficult to cope with the power supplies for both Eurostar and domestic services. Higher speeds are possible with electrification than with diesel trains. For example, in the UK the highest speeds of diesels have been around 125 mph although some have done slightly more than this. In contrast, the highest speeds for electric trains have been much greater than this with for example the Eurostar reaching 186 mph and in 2007 the French broke the speed record for electric trains with a speed of over 300 mph. In 2016, a new speed record of around 360 mph was recorded.

Should Railways Modify Existing Track or Build Completely New Track?

One of the dilemmas is whether to have completely new track which might have been cheaper in the long run with the West Coast main line, or to modify existing track. The French have tended generally to provide completely new track for their faster TGV services rather than modifying existing systems. This is partly because a considerable part of France is very flat. The advantages of completely new track are that much higher speeds are possible with segregation of high-speed trains from the ordinary services. On the other hand, the costs are not spread across different types of traffic, but the TGVs have generated enough traffic to offset this disadvantage. Currently the French are extending their TGV system in France and to neighbouring countries, such as Belgium, Luxembourg, Germany, Switzerland, Italy, Spain and Netherlands. Japan with its Shinkansen (Bullet) trains also has dedicated track.

Legislation About the Railways in the UK

The Railways act 1993 separated the management of rail infrastructure from the operation of trains.

The DfT controls the railway network and is meant to ensure that the right number of trains run and that they operate safely. The rail franchise operators obtained £8 billion from passengers in 2013–2014 which is double the amount that it received at the time of rail privatisation.

The rail executive procures most of the passenger rail services. It is often said that private companies submit bids for the passenger services. However, SNCF the French nationalised industry, runs some services. Spoorwegen the Dutch nationalised rail company runs Abellio which runs the Scotrail franchise, and MTR is part owned by the Chinese government, and currently will be running Crossrail now renamed the Elizabeth line services.

Network rail is responsible for the infrastructure including signals, bridges and many stations. It charges train operators for access to the rail infrastructure. There are five yearly periodic reviews which the Office of Rail and Road (ORR) oversees. The problem with five-year reviews is that it does not necessarily coincide with the period during which improvements, sometimes called enhancements, take place.

Network rail is a not-for-profit company limited by guarantee. Its debt is part of the public sector borrowing requirement. Performance is judged by the percentage of trains that arrive at the station in time and also by the percentage of cancellations.

One of the criticisms, however, is that whilst punctuality is a desirable characteristic it is not helpful if passengers can see a train departing which would connect with their service Sir Roy McNulty in 2011, who had been commissioned by the then Labour government under Prime Minister Gordon Brown, published his report on the rail industry to the then coalition government under the then Prime Minister David Cameron.

Amongst his findings, he stated that there was fragmentation of operations and that efficiency savings could deliver cost savings of £700 million to £1 billion by 2019. Transport Focus represents both customers for road and rail following the Infrastructure Act 2015.

Length of Railway Franchises

There have been disputes about how long franchises should be. Train operators would suggest that they need long-term typically 15 to 20 years for a franchise so that they can generate income through improvements. A longer time period would enable them to plan more thoroughly including possibly the effects of electrification.

Heritage Railways

There are also heritage railways, and currently there are about 100 heritage lines which employ around 3,700 staff where it is estimated that they are worth about £250 million to the UK economy each year. Sometimes these services are complementary to the conventional system. The Bluebell line gives passengers connections at East Grinstead. The Swanage line gives a connection at Wareham to the main line services to Waterloo. In Yorkshire, the Keighley and Worth railway could be used for commuting.

Major Rail Freight Operators in the UK

There are five major rail freight operators in the UK. One of these is a subsidiary of Deutscher Bahn.

Freightliner moves containers from ports to rail freight terminals and was bought by a US firm Genesee and Wyoming in 2015.

Direct rail services are a subsidiary of the nuclear decommissioning authority whose main business has been transporting spent nuclear power to Sellafield in Cumbria.

GB Railfreight is a subsidiary of Eurotunnel. Currently coal movement has been the main commodity which is being moved but this will fall as the climate change convention takes effect. However, transport of biomass is becoming more important and some operators will gain since transport from lorries could move by rail especially when HS2 comes into operation.

The Northern Irish Railway System

Apart from the stations in Great Britain there are also over 50 stations in Northern Ireland which has a different 5 ft. 3″ gauge.

The Northern Ireland system is linked to the Republic of Ireland's railway system. There have been train services between Dublin and Belfast, but it is unclear following the referendum whether passport checks will be necessary at the border.

The EU has been important and the first railway package came into effect in Britain in 2005.

Brief History of Air Transport

Air transport is the newest mode of transport since although air balloons existed in the eighteenth century, the first regular airlines date from just after the First World War in the 1919–1920 period. It has developed rapidly and is widely used for both passengers and freight.

The Way for Air Transport

The way is costless apart from the provision of navigational aids. Aircraft are useful for overcoming geographical barriers such as seas, mountains, deserts and forests. Economists can compare therefore the non-existent track costs for air with rail or road. This is extremely important particularly in countries where there is little transport infrastructure.

In parts of Brazil, some of the inhabitants will never have seen a bus though aircraft will be a familiar sight. Aircraft may therefore be used where the terrain is difficult and traffic is sparse. This may be more cost effective than building costly rail track or roads, which would be infrequently used.

Optimum Height for Aircraft

For most aircraft, there will be an optimum height at which to fly and therefore airline operators will try to make the best use of this subject to safety regulations. For instance, the Nigerian Government gave in the early 1970s fuel saving bonuses to pilots to ensure that they did fly at this optimum height.

As far as possible aircraft will be allocated to airways which may be up to 120 miles wide and 1,000 ft. deep to avoid collisions. Following aircraft in the same airway may be at 100 mile intervals depending on the route. This means that if there are several aircraft, the ones on the outside will do considerably more mileage. The greater the accuracy with which aircraft can be located the smaller other things being equal, will be the size of the airway.

Aircraft Flying Over Other Countries and Noise Problem

One of the problems with air transport is aircraft flying over other countries. One of the problems facing Concorde the supersonic jet was that with the high noise levels it was felt originally that some of the West African countries and the US might not allow such a plane over their territory. For political or other reasons, it was not always possible to fly across the former USSR and therefore aircraft did not always take the quickest route. Currently there are problems of flying between the different parts of Ukraine.

Air and Other Modes of Transport Compared

Countries which do not wish to have traffic from abroad are not, however, confined to air traffic. The problems are possibly less acute with air as whilst diversions are annoying they add to the variable costs in the short run but not generally to the fixed costs. This would be in contrast to the railways where the Tanzam railway was built between Tanzania and Zambia partly because of the problems that landlocked countries wished to transport goods but did not wish to be reliant on sending goods to and from South Africa during the apartheid era but preferred to send them via other countries.

Railways that had been built to run across Germany had large hold ups at frontiers when the country was divided between East and West Germany.

Similarly, railways that were built in India before 1947 and Indian independence had more problems after 1947 when Pakistan became a separate country and in 1970 when Bangladesh separated from Pakistan.

Range of Aircraft and Importance

The range of the aircraft may also determine the route. In the early days of air transport, the polar routes were not always used and some of the earlier airlines when crossing the Atlantic would have to make a stop in Ireland en route.

Air and Rail Compared

The advantage for airlines is that it is possible to overtake other aircraft whereas with other modes of transport, particularly railways, this can only be done if there is provision for passing loops or possibly additional platforms at stations. The other possibility is that the traditional mainlines in the UK where there were often four running lines, the outer two of which were the slow lines with the inner two being for the faster trains. It also means there is not the need for aeroplanes to run at the same speed in order to maximise capacity. This is not, however, true with rail traffic or with traffic on the motorways.

Navigational Aids

The use of radar has become widespread since the Second World War. Aircraft have their own advanced radar systems. This is sometimes called fly by wire. Even if aeroplane crew know where other aircraft are, there still have to be rules to guide them as to what actions they should take.

Therefore, we have international conventions and regulations. If we compare this with other modes of transport it shows clearly there are procedures laid down by the government about where and when to turn. In air transport where almost by definition many flights are international, international rules are needed.

Height of Aircraft

Different aircraft fly at different heights for fuel economy reasons (e.g. turbo props fly at 15,000 to 25,000 ft.), whereas supersonic aircraft may fly at about 60,000 ft. Jet aircraft will find that their optimum height will increase as they continue during their journey as the fuel is used up.

Stacking Areas

When congestion occurs at airports, aircraft are held in stacking areas some distance away from the airports. In these aircraft are kept at a constant height usually at about 1,000 ft. intervals and they gradually descend as the lowest aircraft comes into land.

Aircraft, which run out of fuel, will be diverted to other areas. The problem of stacking cannot be blamed entirely on lack of airport facilities. Airlines do not help (especially perhaps in the USA) where they often reschedule times. Paul M Danforth (1970) in his book "Transport Control" states that at Chicago's O'Hare Airport, 200 departures were scheduled for popular periods from 6:00pm to 8:00pm and 18 of these were at 6:00pm itself.

Noise on Landing

The Department of Trade and Industry publication "Action against aircraft noise", progress report 1973 commented that whilst many people consider that aircraft come in too low the need to follow radio beams which do not bend means that usually have to follow an angle of not more than 30°.

Aircraft Noise and How Measured

One of the main aviation problems has been noise. This is usually measured in decibels PNaB and these measurements are on a logarithmic basis (the threshold of pain is reached at about 150 decibels, a freight train travelling at 40 mph would be about 98 decibels).

A 10 PNaB change would double the annoyance caused to people. Whereas the DC8s and 707s were very noisy aircraft, newer aircraft such as the Lockheed Tristar were far better. Noise increases more than proportionately to speed so that speeds have sometimes been restricted shortly after the take off period (at about 1,000 ft.) though some pilots have suggested that this will impair safety. At Gatwick, trees have been planted in an effort to minimise the effects of noise. At Heathrow in 1972 the noise limits were 110 PNaB in the day time and 102 PNaB at night time (reflecting the point that noise is more objectionable during the evening), night is usually defined as meaning between 23:30 and 06:00 hours. Currently, the government now uses LEQ contours to denote the community response to noise disturbance with 57 LEQ representing the onset of disturbance 63 LEQ representing moderate disturbance and 69 LEQ representing high disturbance. The number of people within the 69 band around Heathrow fell by 85% between 1990 and 2005.

Minimum Noise Routes

These have been used in order to minimise noise over residential areas and are not always the shortest routes. Whilst the Air Traffic controllers will try to keep aircraft to these 10-mile wide airways this may not always be possible because of such factors as wind direction and safety. The centres of such air corridors will be marked by navigational aids.

Noise continues to be a problem and is one of the reasons why in 2016 the government had postponed decisions about the expansion of Stansted and Heathrow. It later stated that it favoured the expansion of Heathrow.

Direction of Runways

Aircraft cannot usually take off and land with a strong following wind but subject to this limitation, most airports will try to ensure that they take off over open country or the sea rather than over urban areas.

Easier Access for Passengers

Especially for short journeys the faster speeds of aircraft have often been nullified by the time, which people have to report before departure of the aircraft. British Airways ran a walk on service with shuttle services between London – Glasgow as from 1975 and then additionally Edinburgh, Paris, Amsterdam, Brussels and Dublin. There was no need to reserve seats which might be brought at the door or possibly even on the aircraft and they claimed that there was no question of passengers not getting seats because there would be sufficient aircraft to cope with peak demands. Currently delays are common for security reasons. There have also been complaints about overbooking of passenger services and passengers being removed from flights when they have boarded the plane.

Air Accidents

It is difficult to assess the improved safety of aircraft. There are many different ways of measuring it, including the number of people killed or injured per passenger kilometre or per hour travelled. This is not entirely satisfactory since sometimes aircraft can kill other people as with the Shoreham air show disaster in 2015 when people on the A27 road were killed. However in 2017 there were no collisions of commercial aircraft.

Perceptions of risk are important to the industry and people are more aware of this factor for aircraft than other modes of transport since there are often a large number of people killed in single accidents whereas the number of people killed per day (about 4) in the UK (in the 1970s about 20) on the roads attracts little or no attention except if it is a celebrity such as the late Princess Diana (1961 to 1997). Road accidents in the USA are much higher proportionately.

Precision Approach Radar (PAR)

In this system, the precision talk down controller takes over from the radar system when the aircraft is about 8 miles from runway and at a height of 2,000 ft. and then issues instructions by radio to ensure that aircraft come down at the right rate of descent and they can then come down visually. It is more flexible than the instrument landing system (ILS), since with the ILS systems the pilot has to have a fixed path approach.

National Air Traffic Services (NATS)

This is a joint function in the UK originally provided by the Ministry of Defence and the Civil Aviation Authority, which was set up in the early 1970s. It provided these air traffic services at the British Airports Authority airports as well as some municipal ones such as Liverpool (Speke Airport) and Birmingham. At Heathrow Airport, it directs aircraft which fly in specific airlines until they can be taken over by Heathrow Airport approach control. It is now a joint public private organisation, which provides control not only over the UK but also over the Eastern part of the Atlantic.

The Canal and Inland Waterways System Land Requirements

Canals ideally need flat land. Often in the UK the railway lines follow the old canal routes for the same reasons. Whereas some commentators have suggested that there is a need for a 300 ft. contour canal system since this would minimise cost, this would not serve most of the major centres of the UK population. It is possible through either individual locks or sometimes flights of locks as with Bingley in Yorkshire to have canals going from one level to another. However, the disadvantage is that it slows down both commercial and leisure traffic. The number of locks meant the canals lost a great deal of their traffic to the railways in the nineteenth century. There are some alternatives such as having cranes, which could haul the boats from one level to another. The most ingenious is the Falkirk Wheel in Scotland, which transports the boat plus the area in which it is sitting in from one level to another, with a minimum amount of energy claimed to be only that of boiling a kettle. This links different parts of the canal system in Scotland. There have also been boat lifts in other parts of the canal system. The choice is not always between inland natural waterways or canals since from the sixteenth century onwards with the river Exe in Exeter there have been canalised rivers.

Inland Waterways and Commercial Traffic

Rivers themselves historically were used a great deal for commercial traffic. In Europe both the Danube and the Rhine are still important. There are however, problems when water supply has become more problematic probably because of climate change and this was dramatically shown in the early 2000s when the Danube dried up leaving many vessels stranded.

Both rivers and canals may need dredging. This will add to the costs. Sometimes the original ports have been so silted up that they can no longer be used for their original purpose. There can be problems of wake particularly with canals so that higher speeds will mean more cost for the inland water authorities.

Waterway owners often therefore impose speed limits. The inland waterways whether canal, rivers or estuaries often have few controls over the craft that can use them.

Ports

There is no satisfactory definition of a port. Landing space for ships can vary tremendously. With hovercraft, it can be a flat beach whereas at the other end of the scale ports can have very expensive equipment. What is necessary is usually flat land so that cargos can be easily transported from ship to shore.

As the size of ships gets greater, often the location of the ports has also changed. This can be seen with the River Thames where the original location in the East of London was no longer satisfactory for many purposes and the port therefore moved away to Tilbury which in the 1970s employed about 20,000 people. Since then the number of people has fallen considerably as Thamesport on the Isle of Grain in Kent and Thames Haven in Essex have become more important.

Few ports can accommodate the largest ships. However, smaller craft only need small scale wharfs and these are available in many areas and are still widely used in spite of the larger ports. Often accessibility time will be lower than with the major ports.

The relative importance of different ports has changed over generations. Whereas traditionally the West Coast ports such as Liverpool and Glasgow were predominant, as more trade took place within the EU, the East Coast ports have become more dominant. It is unclear how the British exit from the EU will affect trade.

In Bristol, which had been a major port since medieval times recently Portbury has become important. As international freight traffic becomes more important as well as short sea traffic to many parts of Europe, the accessibility time becomes more important. Until the container revolution ships often spent large percentages of their time in port even on the largest journeys.

Sometimes as with the boat trains, which took trains on board from Dover to France accessibility was easy and passenger's luggage could remain undisturbed. The Channel Tunnel opened in 1994, has the same advantage.

There have been problems with rail accessibility to the South Coast ports so that Freightliner traffic is unable to reach any of the channel ports except for Southampton.

Roads

Whereas tarmacked roads have been the norm in the UK and for most Western countries, this is untrue in many developing countries.

Road Intersections Compared with Railways

With the railways, there are problems of intersections. One of the advantages of the road system is that overtaking is much easier than when compared with the railways. Even with single track roads, it is comparatively easy and cheap to have passing points unless the terrain is very difficult indeed. Because the roads have been usually provided by the government, relatively little attention has been paid to the cost of rural roads compared with the few users, and therefore the cost per user can be high and some transport economists have suggested that such roads could be closed.

The problems of intersections are important, since most accidents occur at intersections. They are likely to cause hold-ups. Traffic lights can be used at intersections, which can be costly to provide and use some electricity. Traffic lights can be useful as a measure of control if people are making mainly linear journeys since they help to enforce rules about driving speeds. By gearing speeds to changes of traffic lights, people have less temptation to break the speed limits, because if they go too fast at one junction they will merely be held up at another. Traffic lights can be annoying in very off-peak times when people are held up, but on the other hand may be helpful when traffic flows are reasonably even in both directions. One problem, however, is that often traffic lights are not geared to the needs of pedestrians so that if there is not a separate provision for these, pedestrians can be held up for a long while, or alternatively may be tempted to take risks. In the UK, the green man symbol is used to denote when pedestrians can cross the roads. Some of these in London show how long they are going to remain green. Older people have often complained that the time allowed is insufficient and it can cause accidents. Roundabouts take up more land, which may be expensive.

Grade separated crossings can be made, but these are very expensive. There are examples such as Spaghetti Junction in Birmingham. This is helpful since clearly high-speed crashes are likely to be extremely dangerous.

Segregation of different types of traffic were identified very strongly in the Buchanan report more formally known as Traffic in Towns in 1963. Sometimes as in Belgium, there are separate roads, which are used solely by cycles and mopeds. The advantages of mopeds from the countries' point of view is that they use less fuel and also cause less congestion than if people were to travel in cars. On the other hand, mixing mopeds with cars is undesirable since the death rates for motorcyclists generally are extremely high as figures from Transport Trends or Transport Statistics.

Provision for Cyclists and Pedestrians

In the UK, even where cycleways have been designated, they are often just merely another part of the road network and parked cars will obstruct them.

Also without a separate barrier many cyclists will feel unsafe. Pedestrian provision has often been poor in both urban and rural areas. "Brake" a charity, trying to improve safety has an excellent website showing what can be done. There have been examples such as escalators in Barcelona between one level of streets and another which help. Subways are often unattractive to pedestrians since they feel vulnerable especially in the evenings. The possibility of footbridges as on the North Circular road in London reduces accidents. Sometimes the answer might be to have the pedestrians separate as with the shopping malls in Bath. These are still very popular 200 years after they were created. Pedestrians can both shop in safety and are free from pollution and noise.

There are more pedestrian only precincts in the UK and other countries sometimes for the entire week and sometimes just on market days. If it is not too costly at junctions it would be better if road vehicles had to dip rather than pedestrians.

CHAPTER 11
Shipping, Ports and Inland Waterways

Shipping is generally the business of transporting goods and people in vessels over the surface water. The selection of vessels is important to shipping operators and major technological changes have affected its operation.

Measurements of Capacity Including Tonnage

There are a number of different ways of measuring capacity which include deadweight and gross tonnage. Gross tonnage is often erroneously still called gross registered tonnage which is a measure of the internal volume of the ship. Net registered tonnage is a measure of the space, which is available for the carriage of either goods or passengers. Lightship is the name given to the total weight of the ship itself. Deadweight is the total weight of the ship in a laden condition minus the lightship weight. It therefore includes the weight of the crew, passengers, fuel, water and stores. Container ships are generally measured in terms of the standard 20 ft. long containers that they can take. Roll-on/roll-off ferries are generally measured in cargo lengths (i.e. the number of metres by number of lanes), although the width per lane may vary slightly from ship to ship.

Larger Vessels

There is a tendency towards larger vessels. The rate of increase in the size of vessels and total world fleet size has been great. In 1956, the total world fleet was 86 million tons (gross tonnage), and in 1962, the total world fleet increased to 121 million tons (gross tonnage). The world merchant fleet stood at 895.8 million deadweight tons (dwt) as on 1st January 2005, representing a 4.5 per cent increase over the start of 2004. This increase in size has mostly been in the bulk trade, including tankers, containerships and other specialist bulk carriers. The world merchant fleet in 2014 was 1,166,459,000 tons according to Equasis.

Tankers

These are very large vessels used in the transportation bulk liquid cargo (e.g. crude oil). In 1956, the biggest tanker was about 50,000 tons dead weight whereas today 500,000 tons deadweight is common. The rate of discharge of cargo is faster with bulk carriers. The large size has several side effects. Special handling equipment is required at the terminals. The large size requires deeper water and only a few docks can take them. Insurance costs have increased on the large vessels. Since few ports have facilities for their repairs, they may have to be towed for very long distances to any of the few ports that can accommodate them for repairs such as in Japan.

Cellularships

These are large specialist vessels designed to carry containers. The holes are partitioned into hives by steel structures and containers are dropped between the spaces. They are prevented from swaying by the steel structure. Special handling equipment is required and provision must be made for the immediate distribution of the massive size of cargo to be discharged. Computers are used in the loading and discharging of cellularships.

Roll-On/Roll-Off

This was used during the Second World War by the US to load and discharge military equipment and men during a campaign. The idea has since then been commercialised. Cars and lorries instead of military tanks run into the ship and are also driven out. Shipping operators like roll-on roll-off ferries. They are used very considerably on the cross-channel routes from the UK for both holidaymakers and commercial traffic. The largest roll on/roll off ships are probably those between the US and Puerto Rico.

Advantages

They use less terminal facilities than most other vessels. There is little cargo handling, reducing the turnaround time of the vessel. They are used in the transportation of other types of cargo. Palletised cargo can be loaded in roll-on/roll-off vessels by forklift trucks driving on to the vessels.

Containers on wheels are driven in to the vessel and simply driven out at the other end of the voyage. This means that virtually no handling equipment is required. It can be used for trade at ports in developing countries.

Disadvantages

However, more space is allowed for the movement of chassis units and forklift trucks. It requires reinforcement in the strength of the tween decks to be able to bear the weight and impact on the movement of trucks and heavy vehicles. It adds to the fixed cost of the ship but reduces the variable costs such as handling.

OBO Ships

The all-purpose vessel can carry ores, chemicals, packaged timber in shipload. They are normally classed as bulk carriers.

Advantages

They can be used for more than one type of cargo. While specialist tankers do have problems of back-load, the OBO ships can carry other shiploads on return journeys.

Lighter Aboard Ship (LASH)

This is a specialised vessel built to carry barges. Lighters sail from rivers to the coast where the ship is waiting. The lighter is lifted on board the ship with its content.

Loading time is reduced and handling costs cut down. Goods can be loaded at the premises of the consignor or shipper in barges and then transported down the river and out to the sea where the LASH awaits the barges. No port is required since the mother ship can wait in sheltered water.

Disadvantages

Very heavy handling equipment is being carried which reduces the payload of the vessel.

Break-Bulk Ships

These are the conventional ships with derricks and gear for handling cargoes individually or in smaller units. Despite the enormous advantages of specialised ships and bulk carriers, conventional ships still operate in the shipping trade.

Many parts of the world do not have adequate port facilities to handle bulk carriers. Some developing countries cannot afford the cost of expensive and sophisticated handling equipment. Sometimes shiploads cannot be found for bulk carriers. Break-bulk ship-owners used their vessels for liner conference trade rather than tramp shipping.

Tramp Shipping

This is the vessel not used for regular services and usually finds employment in homogenous dry-bulk cargo. In the UK, tramp shipping was traditionally mostly in the coastal coal trade. Tramp ship operators charge freight rates according to the economic demand and supply for the services. There is no specific route and schedule for their operation. This leaves the operator to use his vessel, when there is a boom and high demand for his services, and probably to sell the vessel when the business becomes unprofitable. The majority of tramp ships are privately owned. Tramp ships were sometimes hired by members of liner conferences.

Disadvantages

The tramp ship operator must be aware of the present economic trends and be able to forecast for the future, otherwise, if the owner is caught in a period of economic depression, he/she might lose all the profit he/she had already made. The general tendency is that the operator sells the vessel just at the time a depression is about to begin, and buys ships when anticipating a rise in economic activity.

Liner Conferences

Where several shipping companies were operating on a particular route, they formed themselves into a loose organisation in order to provide a regular service and agree with shippers on fixed freight rates. Liner conference rates did not fluctuate much with changes in demand or supply; therefore, the shipper was assured of a relatively stable price.

Advantages to Operators

Operators voluntarily bring themselves together in order to be protected against the pirate operator who might come in good time to "skim the cream off" the trade. They voluntarily agreed to operate on schedule notwithstanding the losses they might incur during the hard times. This they hoped to recover in boom periods. Shippers are required not to change customers unnecessarily to use the ship of a member of a conference. A shipper receives a certain rebate after a period. This gave the conference members some assurance of trade for a time. Members also shared the little traffic available during unfavourable periods.

Advantages to Shippers

The shippers gained by being assured of a regular service for his trade. The fixed rate also saved them from exploitation by tramp operators who would raise their rates when there is a higher demand for ships.

Many developing countries, who now own shipping lines, did join conferences. They will share in the international trade and thereby gain foreign currency for their country (e.g. UKWAL).

UK West African lines consisted of six shipping companies – British and West African. One shipping company could be a member of more than one conference and there could be more than one conference on one trade route.

The liner conference system prevented cutthroat competition between operators. However, operators still competed since the shipper had a variety of ships from which to make a choice. Sometimes shippers formed a Chamber of Commerce as a "Countervailing power". The Chamber of Commerce discussed standard and quality of service, as well as freight rates with conferences, on behalf of shippers.

Disadvantages to Shippers

During a recession period, the prices did not fall, as they would have done with tramp shipping.

Flags of Convenience

Ship owners can register their ships in any country of the world, although the owners may not be nationals of that country.

Tax and Regulations

Methods of taxation differ from one country to another. Ship owners prefer registering their ships in countries where they will pay less tax. Regulations are not as stringent in some countries as in others and therefore operators would register their vessels in countries where statutory regulations for facilities on board a ship may not be as expensive as in others. The ship owner reduces his operating costs. Nations normally associated with low tax are Panama, Liberia, and Honduras, hence the name "Panholib". Most American ships are registered in Liberia.

Cost of Labour

Sometimes cheap labour may be provided in the country where the ship is registered, sometimes because of the low standard of living in that country or weak trade union movement (e.g. India and some of the former British Colonies). However, the International Labour Organisation has its regulations on conditions of service and wages for seamen from all its member countries.

Flags of Discrimination

This is whereby some countries insist that certain cargoes should be carried in ships of their own flag. This method has its advantages and disadvantages.

Advantages

It is used by countries to ensure a minimum amount of traffic for their own shipping fleet. Countries hoped to earn more foreign exchange by this method. Some nations also adopt this practice for military reasons. The then newly independent nations who were protecting their infant industries adopted this practice. Other reasons include (i) national prestige; (ii) balance of payments.

Disadvantages

It has a retaliatory effect. Shipping operators who are likely to lose some of their traffic by this practice will encourage their own governments to adopt the same method against foreign ships.

In order to encourage the growth of the shipping industry, some countries offer government subsidies to operators whose ships are registered under their own flags. This might affect the efficient operation of the service.

Britain and other West European countries abstained from this practice knowing that they are the old established maritime powers and adopting any form of discrimination might have a snowball effect.

Some countries like Brazil, Uruguay, India, Argentina, Indonesia and Pakistan have adopted a method of "cargo sharing". In the USA, 50% of the aid and relief cargoes were reserved for US flag vessels.

Types of Services

Ferry Service

This is usually over short distances (e.g. on the lakes, cross-rivers and waterways). The Tilbury/Gravesend ferry service was originally provided by railways North of the River Thames as a part of the transport services between the two riverbanks.

The English Channel ferry services to the continent serve both passengers and freight. The flow of traffic across the channel has been growing over the past few years, which led to the demand for the Channel Tunnel opening in 1994.

Coastal Shipping and Short Sea Trade

In Great Britain, this service covers the UK ports, Irish Republic, the Baltic and West European countries. Around the UK ports, it is mostly tramp shipping carrying dry bulk cargo. The fall in demand for coal and the improvements in land transport brought about the decline of the coastal services in recent years. Operators in the short sea trade do not have difficulties in obtaining return loads due to natural trade, (manufacturing) among the countries involved, therefore it is attractive to tramp shipping.

The deep-sea trade includes all the other services, except for coastal and short sea. For example, the Far East, North Atlantic, and South Africa, South America, etc. The imbalance in traffic flows is more pronounced in the deep-sea trade than others. Ships carrying manufactured goods in containers to developing countries with mainly primary products have difficulty in obtaining back load for the homeward voyage. Arrangements are therefore necessary to be made with operators to ensure a certain amount of traffic for the ship. Conference Shipping was very strong in this sector. The deep-sea trade was dominated by liner conferences.

Passenger Liners and Cruising Services

Few people now use passenger liners to go to the US from Europe. This is mainly because air transport has become much cheaper. Many potential passengers were concerned about bad weather conditions, but modern ships have stabilisers.

Passenger liner operators have therefore converted their vessels for cruising purposes. Norwegian ships operate cruising services for the very rich and provide luxurious amenities. The Greek ships have tended to be for cheaper travel and provide services for young people who are less well off. In 2006, Royal Caribbean announced that it was ordering the most expensive passenger ship ever at a cost of $1.24 billion, which would be able to hold up to 6,400 passengers when it was launched in 2009. 'Allure of the Seas' is now the most expensive liner at a cost of around $1.5 billion. Passenger liner services, however, had lingered on for a long time because emigrants to Australia would prefer to accompany their household luggage.

Hovercraft

These are amphibious vehicles which can go over any flat surface. The first commercial hovercraft is usually attributed to Sir Christopher Cockerel (1910–1999). They had widely been used in cross channel services although because of rising fuel prices they have largely been superseded by catamarans.

Catamarans

These are twin hull vessels, which are more help in rough weather and rough seas than the hydrofoils which they have largely replaced. The largest high-speed catamarans such as the Stena Voyager go at 40 knots across the Irish Sea. They are also widely used along the Norwegian coast.

Inland Waterways

This form of transport has declined in Great Britain because of due improvements in other forms of transport. Inland Waterways themselves are not usually large enough to allow the use of unit loads, which have tended to increase in size due to the use of larger handling equipment to the UK. In the UK, the rainfall is not sufficient to provide an all-year-round high water level to be used by the river craft. The provision of locks and the pumping of water in and out are very expensive operations.

The Ministry of Transport had classified waterways into two parts (1968 Transport Act): commercial and cruising. This gives rights to interested people to develop some of the country's canals for commercial purposes while others remain for cruising only. On continental Europe and in North America inland waterways are extensively used for commercial purposes. Barges up to 300 tons can travel as far as 100 miles up the River Rhine. In America, 1,000 ton barges are common.

At the peak of the boom in waterways in the UK there were about 4,000 miles of navigable waterways, but in 1974 this figure was only 2,500 miles. About half of the navigable waterways in the UK are owned by British Waterways.

The main waterways which are used for commercial purposes include the rivers Trent, Severn, Lea, Weaver, Aire and Calder Navigation, Sheffield and South Yorkshire Navigation, Gloucester and Sharpness canals and the Scottish Waterways, also the Caledonian and Crinan Canals which are retained for strategic reasons.

This group of waterways carried 56 million ton miles during 1972 in three roughly equal parts (i.e. coal, liquids in bulk and general merchandise).

All but about 400 miles the remaining waterways (mainly canals) are used primarily for amenity purposes. The others, which are classified as remainder waterways, have a more uncertain future but might be used for leisure purposes.

One of the main problems with the waterways in the UK is water supply since the majority of rain falls at the time of least traffic and therefore reservoirs are needed (most of these are over 150 years old). Maintenance of the way is necessary (i.e. there must be inspections of banks, locks, bridges, tunnels). There are also large scale dredging operations.

Advantages

They provide natural ways for transportation. They reduce road congestion on the roads. River craft are less noisy than lorries and other road traffic. More valuable land being used for motorways can be saved by diverting some traffic from the road to inland waterways. Goods can be delivered directly to the premises by the consignee on the water bank.

Less energy is required to move the same amount of cargo by water than by land.

International Maritime Organisation (IMO) – Formerly Inter-Governmental Maritime Consultative Organisation (IMCO)

The United Nations Organisation set up this agency in 1948 at the UN Maritime Conference held in Geneva. Its purpose is broadly to provide smooth operation of shipping in international waters. It will consult and advise governments on matters such as navigation at sea, pollution, safety, and the provision of an international trade, which is as free as possible from restrictions and discrimination.

The membership had increased to about 160 since its formation in 1948. Various committees are set up to look at the different sectors affecting international shipping (e.g. The Legal Committee, Maritime Safety Committee and Technical Co-operation Committee). There are several other sub-committees dealing with various subjects.

Marine Pollution

IMO is very concerned about pollution of the sea by large oil tankers. Limitations are being placed on the amount of oil that can be discharged by an oil tanker into the sea. Other measures include the inspection of ships at foreign ports, detection of offences and penalties in respect of unlawful discharges of oil and other harmful substances.

Routeing of Ships

Regulations are made for the prevention of collisions at sea. Basic principles to be observed by ships when navigating are set out to ensure that adequate navigational aids are provided.

Life Saving Appliances

Up to date requirements for life-saving appliances are spelled out, such as inflatable rafts, markings, lifesaving appliances, etc. There have been concerns during the so-called migration crisis in 2017 that many refugees are using unsuitable boats and many people have been drowned trying to cross the Mediterranean.

Search and Rescue

International plans for search and rescue operations are drafted by the organisation. SAR Convention and its technical provision are intended to be applicable in all countries.

Standardisation of cargo handling equipment is also reviewed by the organisation. Conditions are set for the carriage of dangerous goods, hence the IMO International Dangerous Goods Code.

Legal Work

Shipping by the nature of its operation is international. Laws differ in different countries. Legal matters in those areas affecting shipping are discussed and general standards are set by IMO. IMO also regulates workings on board the ship including passenger and crew safety.

Boat Trains

Sometimes as with the boat trains which used to take trains on board from Dover to France accessibility was easy and passenger's luggage could remain undisturbed, but this advantage was not really reached again until the Channel Tunnel opened.

The most famous boat train was the Orient Express which originally ran from London Victoria to Constantinople which was then the capital of Turkey.

Freightliner Traffic

There have been problems with rail accessibility to the south coast ports so that Freightliner traffic is unable to reach any of the channel ports except for Southampton.

CHAPTER 12
Airports

The aviation industry is extremely important in the UK, partly because of its effect on the UK balance of payments. The majority of tourists who come to the UK do so by air, and the tourist industry is itself a major component of the invisible part of the balance of payments.

Multiplier Effect of Airports

Whilst commentators have often suggested about the multiplier effect of airports, this is only helpful to the country as a whole if airports led to greater economic activity overall rather than merely displacing economic activity from one area to another. This type of debate is not confined to airports. There were suggestions that the Severn Bridge opening, whilst built partly to help develop South Wales, boosted the Bristol area more leaving the depressed parts of South Wales in much the same economic state. The multiplier effect of new airports can be very substantial with the jobs not only at the airport but also from other activities which tend to cluster around airports, such as the increasing use of new warehouses. The growth of passenger numbers has been very substantial and it is one of the many reasons why the Roskill commission 1971 was looking at airport expansion in the early 1970s. It is not only, however, airports in developed countries such as the US which are very important, but also in many parts of the developing countries including increasingly many parts of North Africa as well as East Africa. This is partly because some countries such as India and China are developing fast and also generating considerable volumes of traffic from those countries. However, as people have become richer, air fares have fallen in real terms and journeys have become faster that people choose a wider variety of destinations than in the 1960s when generally people would fly to other parts of Europe from the UK rather than going to any parts of Africa.

In some countries where the population is relatively sparse as in Canada and even in Australia, internal flights are very important. Whereas before

1939 airport development mainly catered for business passengers and a small amount of freight this is no longer true. The aviation industry has to take into account of increasing criticism of its impact upon the environment as well as the accessibility to and from the airports and its impact generally upon land use planning. There are currently concerns about how the airports can reduce carbon emissions to help comply with climate change conventions.

Business and Social Traffic Compared

The increasing use of chartered flights and budget airlines means that more people have flown, although the total numbers of the people who have flown is still comparatively small. The advent of packaged holidays from 1960s onwards coupled with the abolition of any exchange controls after the Conservative government took power in 1979, also led to the development of more air traffic.

When forecasting demand for airports, economists need to take account not only of the average disposable income, since this may mask great inequalities but also the number of people in the different income brackets. This is because demand for air transport is still very income elastic.

One of the problems for airport operators is how they can accommodate extra passenger growth, but generally with fewer aircraft than was originally expected. However, the credit crunch in 2007 onwards and also likely increases in fuel prices in the longer term means that it may be important to have another look at forecasts especially as many major firms including Alitalia have run into difficulty.

The airline industry has peak periods, which can be seasonal, weekly and daily. Airport capacity takes account of the main runways as well as taxiing runways and terminal facilities.

Congestion at airports can be rife and too often the aviation industry seems to have been more pre-occupied with the vehicles (i.e. the aircraft rather than with the overall time taken). We can see this readily with Concorde, which was built largely with government money, which was a very regressive form of subsidy and where whilst the flight itself might have taken 3.5 hours to cross the Atlantic, passengers could then spend 1 hour recovering their luggage at London airports.

Within the UK, Heathrow has often billed itself as the number 1 International Airport, judging by the number of passengers who go through it. However, whether this means it is popular is more debatable. Heathrow itself has suffered considerably from the noise problems, which are created both from the aircraft themselves, but also from the increasing volume of car traffic, which serves it. Gatwick for a long while only had one runway, but had the advantage of being readily accessible from London Victoria via the Gatwick Express and other railway services. Stansted has much less of a noise problem, since it is in a relatively sparsely populated area. It also has the advantage that British Airports Authority as it then was, bought a lot of the land around the airport so that opposition on grounds of noise has tended to reduce. Stansted also has the advantage of an express rail service from Liverpool Street which itself is one of the busiest railway terminals in the UK. Stansted also has connections to other parts of the country. There are also other airports around the London area such as Luton and Southend-on-Sea. Under the 1978 plans, there were to be no further airports to be built. Since then City Airport in the docklands has come into being. In 2008, the then new mayor of London, Boris Johnson (Foreign Secretary until 9th July 2018) suggested that there should be a new airport near Sheppey in Kent and this hub airport would replace both Stansted and Heathrow.

Lack of Investment Appraisal

Airports have often been provided by the government and have often not had the type of investment appraisal, which transport economists might expect for such major investments.

Roskill Commission

The government in 1968 announced the appointment of an independent committee, to look at the possibility of a third major airport in the London area which came down in favour of Cublington in Buckinghamshire about 7 miles North of Aylesbury as the cheapest site, although a minority of the committee favoured Foulness sometimes known as Maplin.

The commission itself looked at all suitable areas as it saw it within a range of about 50 to 130 km of London, but ruled out most of the larger towns if they were going to be subject to excessive noise. It is not clear however, why this should have meant that Foulness was considered and subsequently chosen, although later abandoned, since it is very near to Southend-on-Sea which has a population of over 120 thousand people.

After the Roskill Commission in 1971 Mr. Peter Walker the then Secretary of State for the Department of the Environment stated that Foulness (Maplin) had been chosen on 5 main environmental grounds. These were:

- It would reduce the noise at Heathrow and Gatwick.
- There was a need for more terminal facilities.
- It would reduce road congestion around Heathrow.
- It would lead to the closure of Stansted, Luton and Southend airports.
- It could be combined with a new large port for supertankers.

This proposal was shelved after 1973 oil crisis. Another Government Minister said at a meeting in Luton that the project would not be so much an airport plus a port as a port plus an airport presumably implying that the port was more important than the airport.

The Port of London Authority (PLA) had already extended its territory to include Foulness some years previously, presumably in anticipation of building a port there.

Air and Other Developments Since the Roskill Commission

There have been a number of developments since the time of the Roskill Commission which makes the above arguments seem more dubious.

Noise

Newer aircraft such as the Lockheed Tristar were much quieter than the DC-8s or the Boeing 707s.

The government if it wished to cut out aircraft noise could do it more effectively and cheaply by pricing out noisier aircraft.

The late Sir Peter Masefield, a former chairman of the then British Airports Authority put forward the interesting viewpoint at the time that since towns are noisy anyway, noise of airports in towns is probably more tolerable than noise in the country.

Professor Flowerdew's Criticisms of Noise Argument

It is interesting to note that though the noise argument was used the government did not try to quantify this argument and Professor Flowerdew, a member of the Roskill Commission, subsequently published his own figures.

Congestion Around Heathrow

Sir Peter Masefield argued that whilst there is congestion around Heathrow it would be cheaper to cope with it there than around the links at the East side of London to Foulness.

The then new Piccadilly extension to Heathrow which is capable of carrying up to 25,000 people per hour would help to relieve congestion, although a better solution would be for a new nonstop railway to run direct from Heathrow to Victoria. This would only need about 3 miles of new line, since it could then come on to existing railway lines. It is interesting to note that over 30 years later BAA has come up with a similar solution but to link to Waterloo.

The 50 million passengers per year it was estimated would use Foulness by the mid-1990s would have generally a much longer journey and this would be a waste of fuel, which is likely to become a scarce commodity.

More Terminals

Whilst the Roskill Commission accepted this argument, there have been several developments, one of which is the growth of even larger aircraft than the Commission anticipated. For example, the average number of passengers per plane was about 180 by 1985 compared with their estimate of 162. There will be (if this estimate by the late Professor Walters is correct) about 10% less flights than if the original estimate was correct. The cost of fuel would in any case probably lead to even larger aircraft being used.

There have also been improved techniques of air traffic control and the likely growth of fuel restrictions (partly resulting from the Arab Israeli War in November 1973 which led to major airlines reducing the number of flights across the North Atlantic) meant that both Luton and Southend could be used to help out.

There is perhaps a need for extra terminal space, but the government had stopped the building of these at Heathrow presumably to build up public pressure for a new airport at Maplin. It is interesting to note that more than 30 years after this the fifth terminal has finally been allowed.

The cost of building the extra terminals had risen enormously so that by November 1973 it was estimated that the cost of building a two runway airport (at 1972 prices) compared with about £250 million to build the same facilities at existing airports.

The Channel Tunnel project went ahead and this took away some of the short distance traffic (the then current estimates were that it would take away 6 million more people than the Roskill Commission estimated).

Essex was also to have heavy industry but the government policy by late 1973 was that the area was suitable for light industry. The analogies which had been drawn with Rotterdam which was generally regarded as an efficient port seemed odd as Rotterdam was not situated at the mouth of the river.

Consultation with the People Affected

The Department of the Environment drew up a list of 6 possible routes, which could be used for Foulness and asked householders for their opinions. The response was poor perhaps because the Department did not allow people to indicate whether or not they approved of the project. A voluntary body, the Defenders of Essex also sent out a questionnaire, which had a much larger response.

It is perhaps interesting to note the replies of the 12,800 people who said they were against the project.

- 31% thought there had been insufficient research.
- 23% had doubts because of opposition by the British Airports Authority (BAA), etc.
- 9% because it would not really give noise relief from Heathrow and Gatwick.
- 18% because it would not help regional planning.
- 5% because of the fuel crisis.

- 14% because of danger to wild life.

This questionnaire can be criticised on the grounds that people who might have criticised the airport on other grounds (e.g. destruction of housing for access routes) could not have given this answer.

When looking at airport transport, economists need to look at the origin destination matrix. It is usually assumed with the UK that people going to London will then go on to London for business and research seems to suggest that this is the case. However, it is not clear that if, for example, the government built an airport elsewhere and then developed that area that this might not change. Some more radical elements such as the Economist magazine have in the past suggested that Parliament should be located in York, which would then mean presumably that many of the civil servants would alter their location. Since business executives often want to visit civil servants particularly for lobbying purposes as well as for other factors, then this might be considered.

Even at the time of the Roskill Commission it was noted that the majority of trips to airports were for leisure purposes rather than for employment purposes. The number of people working at an airport can be quite considerable and at the time of the Roskill Commission Report it was suggested that around 30,000 people would have been employed at Maplin Airport, which would have meant a large increase in the total population of that area. This, however, would not have been as big as the projected rises in the Thames Gateway currently.

Capacity of Airports

When considering future capacity, airports have often looked at this in terms of the length of potential runways. The International Civil Aviation Organisation (ICAO) has suggested different standards.

When considering the amount of land that would be necessary for a new airport, then there are many factors apart from the cost. One of these is the runway length and this would need to be large enough to cope with not merely current planes but also larger ones. The European Airbus as well as the Boeing Dreamliner is larger than previous aircraft and it seems likely that this trend will continue. Even with smaller airports which would not necessarily generally cater for the larger aircraft, sometimes as aircraft get older they are transferred from the major airports to the smaller ones so airport design has to allow for this.

There is also a need to consider the number of runways and the International Civil Aviation Organisation lays down the percentage, which must be able to cater for the existing traffic. Unlike other terminals, runways have to allow for prevailing winds. The runways would not normally be suitable if they had considerable crosswinds.

Increasingly, whilst aircraft have become quieter there have still been more complaints about aircraft noise and it is likely that this will continue. The airport operator would also need to consider the air space, which may be very restricted in some conurbations. There is also the question of safety since near New York there are a large number of aircraft operating on any one day. There are also the problems that military aircraft can be involved in collisions because unlike other commercial aircraft they do not have to give details of their flight plans. There is also the problem that smaller aircraft often referred to as general aviation aircraft can also be involved in collisions. In the UK, there are about 400 different airfields from which smaller aircraft can be taken.

Airport Commission Report 2015

The report states that no new full-length runway has been laid down in the South of England since the 1940s and contrasts this with other countries where more provision has been provided.

The independent airport commission was set up in late 2012 during the Conservative-LibDem coalition period. The report states that whilst London has good transport connections, Heathrow is already operating at capacity. Gatwick is nearing the same point and states passengers on other areas are transferring to either European airports or even sometimes airports in the Middle East. It also states that the entire London system will be full by 2040.

The commission states that a new runway in the South East is needed by 2030. It also states that aviation helps many other stakeholders such as the media, the financial sector in the city of London as well as the universities in the London area.

Location of Airports

The ground has to be reasonably flat if at all possible and clearly there must be no obstacles. Hong Kong Airport is one where aircraft fly near ordinary houses, but this is not a normal example. There is now a new Hong Kong Airport.

Generally, transport economists would not expect to find airports in very mountainous areas although sometimes as near Bergen in Norway, there is very little choice. There should also be freedom from bird strikes if at all possible and this was one of the objections to Foulness where Canadian Geese were not only thought of as undesirable in their own right but also because they posed potential problems to the aircraft. As aircraft become quieter there may be fewer restrictions on the planning for areas around the airport. Usually transport economists find that where there are high noise levels then they would not expect to find very much in the way of commercial activity apart from agriculture whilst in the next zone there may be industry and warehousing.

Runway lengths will depend upon the type of aircraft, although aircraft manufacturers being aware of this will sometimes try to modify their aircraft so that they can use existing airports otherwise they are likely to be limited in terms of total demand. If temperatures are likely to be high, the runway will need to be longer since aircraft produce less thrust.

There is also a need to have adequate taxiways, which is where the aircraft go before they get onto the main runways. There is a need for exit taxiways so that aircraft coming in do not take up scarce main runway capacity. In some major airports, there may also be a need for parallel runways.

There will also need to be land provided for aircraft fuelling and also for passenger and cargo handling. Heathrow and even Copenhagen need vast facilities.

Waiting Time at Airports

One of the problems in many airports is that whilst there is a great emphasis on speed of aircraft the time taken within the airport can well be considerable. Sometimes airports have provided travellators so that people with luggage, etc. can use these rather than having to cart heavy luggage around. It is undesirable that people should walk any great distance to the aircraft and this has become more important because of the potential security risks. Luggage movement is often through conveyor belts, but occasionally porters are used. This adds to the cost of providing the airports. The number of checkpoints available for any airline can be determined partially through queuing theory (i.e. to determine how long people will have to wait). Whilst some airlines have their own separate check-ins, this can lead to longer delays for passengers. There should also be adequate luggage trolleys.

Accessibility Time to the Airports

There are problems of access to the airports. Whilst aircraft themselves are fast quite often the airports themselves are situated well away from the likely points of origin or destination of most travellers. One of the criticisms of some of the low-cost airlines is that their advertisements can be misleading (i.e. they will give the name of the city, which allegedly they serve, but in fact it may be a considerable distance away). Even with the standard airports, Gatwick is 40 km from London and Stansted even further. Whilst a great deal of data is about the numbers of passengers using the airport, in practice passengers account for generally less than half of the volume of people attending the airport, since there are also considerable numbers of workers at the airport as well as friends and relatives.

Heathrow Access

Even on much longer journeys such as that from London to New York the air travel time is only around half of the total time and this could get even worse for many passengers given the security scares which occur. BAA were responsible for the provision of the Heathrow Express which whilst expensive offers a 15 minute service frequency from Paddington. This has subsequently been amended to allow not only a non-stop service, but also a slightly cheaper service, which will stop at some intermediate stations. This is helpful to rail passengers who otherwise were either forced to take a coach from Reading to Heathrow or alternatively to come into Paddington and then to travel back over the route on which they had already travelled. BAA has also announced plans to have a rail link, which would link to the main rail network, rather than having a self-contained service to Paddington.

For pollution reasons and for safety reasons, such as the risk of fire, the airports would much prefer to have electrified rail services coming into the airport than a diesel service. There is a choice between providing the Underground service such as the frequent Piccadilly line which is not very convenient for people carrying luggage, but which gives access and inter-platform access with the District line to a great number of destinations to London or having the dedicated service like the Heathrow Express. Sometimes airlines may want to run their own airport coach links sometimes with check-in facilities. In the UK, a number of them paradoxically have been listed in the national rail timetable such as those from Reading or Woking to Heathrow Airport.

There have been other forms of access to and from airports sometimes including coach services, which have linked airports such as Heathrow and Gatwick, and helicopter services were used linking Heathrow and Gatwick until the early 1980s.

Access can also be by taxi. There are many advantages of this from the airports' point of view. One is although there is need for some parking whilst the taxis put down and take up, there is not the same need for parking space which private cars require. The taxis themselves have the advantage from the passenger's point of view that there are no parking charges when dropped off, although occasionally there may be parking charges for pick-up. They get a direct route to and from the airport. In the UK, however, it is true that by far the majority of workers use private cars, although BAA for other reasons would wish to see more using public transport. Whilst it is sometimes claimed that the government does not have to pay money for people using their private cars, this is not entirely true since part of the justification for the M25 is that a large proportion of the M25 between certain junctions (i.e. those near Heathrow) is used for private cars going to and from Heathrow which would otherwise be unnecessary.

Parking Problems at Airports

Within the airports themselves, car parking space can be major problem. The majority of car parking space is necessary for a short period generally up to about 2.5 to 3 hours.

There is also medium-term parking which is necessary generally for workers. There is also long-term parking, which takes up a considerable amount of the space. How the airports should cover these charges is a matter of dispute.

Because of problems of security especially after the Twin Towers tragedy in 2001, more attention has been paid to problems of passenger luggage and there are now standard search procedures for all baggage. Apart from these risks, there is a standard problem of baggage theft and only a small percentage of this is recovered.

Charging for Airports

The airports, whilst gaining some of their money from what might be thought of as their standard revenue, (i.e. from landing and take-off) also gain money from other aspects. The airport operator when considering landing fees has to consider the cost providing not only the landing areas but also the runways and the taxiways. They must also depend on the country involved in the methods of trying to recover the cost of air traffic control. The airports have typically tried to recover their landing fees by having different charging fees for helicopters which have the advantage of not requiring very much at all in the way of landing space and also then different charges for aircraft. There may also be another variable charge according to the weight of the aircraft. There may also be cheaper rates in the off-peak season, for example, during the winter months.

In the UK, nowadays there is also a levy (air passenger duty) on all passengers flying out of the UK, which is there allegedly for environmental reasons, although it is difficult to see why it should be charged per passenger and not on the aircraft as a whole. This was changed from 2009. There are also charges levied by the airports themselves on passengers during the peak hours for both domestic and international flights, although the amounts can be higher for international flights. The justification for this is presumably that international flight holders will take longer, although this seems rather odd given that one can fly from, for example, Heathrow to Scotland which would be a longer distance than from Heathrow to Paris. The airports themselves may levy charges for aircraft parking although sometimes there is no charge except at Stansted for the first period.

Sources of Finance for Airports

Airports are unusual compared with other terminals since only around 50% of the revenue on a world-wide basis comes from what might be expected to be the major source of income (i.e. charging for landings). It was noticeable in 2007 with the development of the Heathrow 5th terminal that a large amount of revenue was expected to come from the 127 shops that were scheduled to be built. Because the cost of airport provision is so large, the price mechanism itself is very unlikely to be allowed to reign and even in a country such as Japan, which is often thought of as being ultra-capitalist, the local pressure groups have often determined that the expansion of air should be relatively limited. However, it might be noted here that Japan by contrast to other countries, such as the US, has a very highly developed rail system, which gives very fast interlinks between most of the major Japanese cities.

If economists look at revenue then they will see that there are a large number of different revenue aspects, which can be charged. These can include rentals for shops, which for large airports such as Heathrow can be very considerable. In particular, because of duty free sales, people may find in some cases that they can gain goods at relatively cheap prices, although it is sometimes suggested that the price elasticity for goods sold at airports is likely to be less than those for ordinary goods since people on holiday are less likely to be paying attention to prices than, for example, people wandering around computing shops in a shopping centre. There may also be parking charges, which may also account for a fairly large percentage of total revenue especially from long stay car parking passengers. Additionally, there may have been money from charging for taxis and other spaces. Sometimes, as with the railways before privatisation, they may also provide hotels.

Local Authorities and Airports

Historically, before the Second World War, airports were often run by local authorities who ran their airports partly as a matter of local pride and also partly because it gave a boost to tourism which often meant that whether or not the airport made money was less important than this. To some extent these considerations have still taken place since if people read "The Full Story of the Controversial Low-Cost Airline" published in 2007 and written by Siobhan Creaton, one can see that one of the airports in the Republic of Ireland was developed partly for this reason and it kept its airport charges low.

Airports and the Peaks

The airports themselves tend to have pronounced seasonal peaks, particularly during August and therefore if demands are to be met, although this is debatable, the capacity of the terminals has to be considerable.

There are also generally daily peaks with far more people wishing to travel during the morning period. If economists look at the total time of travel to the airports, Eurostar has produced a quite successful advertising campaign which shows what will happen if people travel by air compared with their services from London to Paris or Brussels.

Peak Pricing

Airports need to have a system of peak pricing, since otherwise most airlines will tend to cluster their schedules into the peak whilst leaving the airport quite sparsely used at other times of the day or seasons.

There is a need also to discourage General Aviation at this time and this can be done as well by peak pricing.

Price Discrimination

Historically there were different rates between international and domestic passengers, but this made sense only when different types of aircraft were being used and increasingly we see the same types of aircraft being used for different types of journeys.

Subsidies to Aviation

There is little reason to subsidise domestic journeys, except in particular circumstances possibly such as those in the Highlands and Islands where there are few if any alternatives available. Again, there is no reason why there should be cross subsidies with duty free for example, being used to subsidise air travel. Most people would regard this as a regressive subsidy. Economists will have observed that often historically airports were subsidised, as have airlines indirectly been via subsidies such as that to Concorde. There is no obvious reason why the average poorer taxpayer should subsidise the rich.

Heathrow Congestion

Whilst Heathrow became near to capacity, it sometimes tried to force airlines to use other airports. Many airlines prefer Heathrow basically because it is possible to get interchanges there between different routes. However, there seems to be no reason why routes and slots should not be allocated through a pricing system in the same way that we would expect this to happen in other modes of transport. There is even less reason given that the social cost of air travel is likely to be greater around Heathrow with its large population compared with that of Stansted. The cargo operators have often had night flights when they would be charged less for the aircraft than at other times. Sometimes this will mean that if a factory operates fairly typically a 2:00pm to 10:00pm shift then the products from this would then be able to go by air transport without any great difficulty. The main controversies about airports have been their externalities.

Limitations of Laissez Faire

One of the problems of course of airport policy, is that because the number of airports is likely to be limited by air space, then it is not possible to have a laissez faire regime since if one airport expanded too fast in one direction, this could well be limiting potential flights from another airport. Generally, pressure groups have been against the development of airports. A straightforward referendum basis as has been tried, for example, in Switzerland will probably show that people are more likely to vote against airports than for them. Even in the USA a large proportion of the population have never flown, and in countries such as the UK the number of people who had not flown is still considerable and of those who do probably fly, it is comparatively irregularly compared, for example, with taking cars, trains or even buses.

Evaluation of Noise Problems

Whilst it is possible to evaluate the extent of noise, part of the problems for the aviation industry is that whilst night flights would improve utilisation of airports they clearly have a more detrimental effect on many people's lives than the same amount of noise during the daytime. One possibility, which some airports have used, is to have a rebate for quieter aircraft and computer-generated information can show which aircraft are quieter than others.

Noise and the International Community

We mentioned earlier the problems of economic development and airports. Clearly whilst Britain (apart from the boundary between the Republic of

Ireland and Northern Ireland) has a clearly defined boundary, domestic flights are generally a nuisance only to the UK. In much of continental Europe, however air flights from one country may well have quite a detrimental effect on neighbouring areas, which could be in a different country.

Noise Reduction

One possibility would be as around Heathrow for the BAA to make grants towards insulation of houses and flats. This has been done for a long while. It would seem logical that this should be paid for by the aviation industry rather than by the taxpayer generally. Another possibility would be to move the airports away from conurbations so that there would be less noise. On the other hand, this could be unhelpful for the environment generally since if people take longer to travel to and from the airport, this is unhelpful in its own right. One of the complaints about airlines is that in many cases they suggest the name of the nearest large town, which the aircraft operates even if the distance is quite large. One example of this is Hahn, which is about 2 hours by coach from Frankfurt.

The Aviation Industry and Fuel

The aviation industry is likely to come under increasing pressure to become more fuel efficient partly because the fears of climate change have become much greater, and partly because of the instability of oil producing countries such as Iraq. Fuel costs are a significant part of total operating costs. The ability to save fuel depends partly upon aerodynamics and increasingly therefore aircraft manufacturers will use computerised design to try to minimise the amount of fuel used.

One possibility would be to have alternative sources of fuel. Clearly it is unlikely that many people would wish to have nuclear powered aircraft although there are some nuclear-powered ships.

The first solar powered flight around the world was made in July 2016. Hydrogen would be another possibility and is being considered for cars, but would be more difficult to use in aircraft mainly because of its low density. This would otherwise mean that a large amount of space would need to be used.

The Development of UK Airport Policy

The government had published "a new deal for transport, better for everyone" in 1998, which was the forerunner of an airports policy, which was meant to look at 30 years ahead. In July 2002, the government published a paper 'The future development of air transport', which proposed expansion such as new runways at Heathrow, Stansted, Birmingham and East Midlands and possibly new airports at Cliffe in Kent and possibly one in Bristol. In November 2002, the High Court had said that it was wrong to exclude options for developments of new runways at Gatwick. The government sets out the outputs of UK airports, looking at the number of air transport movements between 1966 and 1997, where it is noticeable that the number approximately tripled (i.e. from 556,000 movements in 1966 to 1,686,000 in 1997). On the other hand, the number of terminal passengers increased from 22 million to around 135 million (i.e. an increase of approximately 6 fold in this period). The difference between these two ratios is of course that aircraft have grown gradually bigger and this provides one of the problems of trying to forecast future growth. It is also very noticeable that whereas some years had very significant growth such as 1978 when passenger numbers grew by 13.5% in other years such as 1991 the air transport movements fell by 10% and terminal passengers by 4% presumably reflecting the impact of the Gulf War.

UK Government Review Backs London Heathrow and London Stansted Airport Strategy

The government suggested that there is a need for more runways at both Heathrow and Stansted. This could have been done at that period by having a second runway at Stansted to be built between 2011 and 2012 and also a third short runway between 2015 and 2020. It does, however, make a point that it would need to take account of meeting air and noise limits.

The government thinks that more runways are necessary. Many commentators might feel that this is not the way forwards and that the need, for example, of short term and longer-term circumstances are not compatible with the idea of climate change. Whilst clearly in the short term, the US run is not really suitable by other forms of transport if as hoped by the existing Russian government that there could be a Bering Strait tunnel which would open up the idea of a long-distance rail route which might be an attraction for some passenger travel and perhaps more importantly would also open the idea of freight from Europe to the US in timings which whilst slow could well be ideal for many west coast US routes.

For short distance, air routes such as those from the UK to France, Germany and the Low countries, the prospects of even more TGV trains in France plus even more fast routes in Germany with ICE trains or the equivalent will mean that for many journeys the overall times between trains and air is similar. Much will therefore depend upon the pricing policy of the railways and the pricing policy of the airlines which in turn will depend partly upon the taxation policy of the governments concerned including the UK government.

Environmental Criticism of Airport Policy

Dr Lucas then a green MEP and later a Westminster MP, says that *"if we are to stave off the worst impacts of climate change fairly we must cut UK emissions by 80 to 90 per cent in the next few decades: that will simply not be possible unless we reduce emissions from flights"*. This would seem to be in line with criticisms that the government is still using a 'predict and provide' policy which is not capable of reducing either the greenhouse effects or for that matter making the best use of fuel. There seems to be no reason why in many cases high-speed rail links would not be both more efficient from the user's point of view as well as providing environmental benefits.

October 2007 Pre-Budget Report

The government announced in this report that in the future the duty would not be on individual passengers but on flights, which would encourage airlines to make better use of planes. Clearly at the time the duty of £10 per economy class ticket is likely to have an effect on low cost airlines, but not upon more expensive airlines. It might have the effect of encouraging higher price airlines to look for better load factors and the use of the internet might mean that lower fares might be available the closer it was to the time of the departure for less full airlines. It was also expected that the tax component for example of going across the Atlantic might increase to between £40 and £80, whilst for Europe business and first-class passengers would be charged £20.

2008 Competition Commission Report on Airports

In 2008, the then Competition Commission recommended that Gatwick should be sold. This was because with BAA in control of Heathrow, Stansted and Gatwick it was felt that there was an unacceptable degree of monopoly power. It also came after there had been considerable criticism of the terminals at Heathrow when there were many delays at terminal five and for quite a while afterwards there were many unfavourable comparisons of the way that it had opened in contrast to the smooth running of St. Pancras after Eurostar had changed its terminal from Waterloo to St. Pancras in 2007.

Part of the problems that the Competition Commission Report identified was that Heathrow is a hub airport and that the commission therefore recommended that there should not be a sale of Heathrow unless selling either Gatwick or Stansted would not achieve more competition. It also followed that capacity would be a limiting factor in the South East unless capacity was increased. There would, therefore, be a need for continuing regulation.

Forecasting of Airport Capacity

Forecasting of airport capacity and planning the right amount of land in the right place with the right number of runways as well as access to and from the airport is important. In the UK, this will become even more important following the announcement by the Conservative Prime Minister, Theresa May, that Heathrow would be expanded in spite of strong opposition from other airports such as Gatwick and also strong opposition from local people in South West London and elsewhere.

This was shown in the by-election where the sitting MP for Richmond Park in 2017 was defeated by the Liberal Democrats partly because of opposition to the airport expansion.

Airport capacity and changes to it also need to take account of carbon emissions and most countries regard this as harmful and have signed up to the Climate Change Convention.

It is important when looking at transport demand including airport demand that they look at the origin and destination of passengers sometimes called the OD matrix. However, capacity is not just taking into account passenger demand but also freight demand.

There can be some new destinations such as Cuba which after the death of Fidel Castro who had been in power for more than 50 years is now a more open society so that more passengers will be encouraged to go to and from that country than in the past.

There may be alternatives to air transport for some passenger journeys as the digital revolution makes it much easier for people to participate in discussions even on global warming or climate change without actually travelling to and from them.

There may also be alternatives to air transport for longer journeys such as those from China to the UK where the first container load trains took weeks arriving from China to Barking in Essex in 2017.

There may also be reductions in the volume of air traffic freight as miniaturisation of many products takes place.

Readers can observe this with the change in computers where originally there were mainframe computers taking up a lot of space but increasingly masses of information which would have occupied a large room can now be held on a memory stick.

Virtual reality often means that people do not have to travel to see architectural plans but can instead use these to be able to discuss details even if afterwards they may want to visit a particular site.

Apart from the externalities caused by air traffic, operators, passengers and other stakeholders will need to forecast the likely change in fuel prices. The difficulties of this can readily be shown by the sudden halving of oil prices in 2015 which few commentators had anticipated.

Newer aircraft will anyway use less fuel whether these are manufactured by Boeing or variations of the European airbus or other possible new consortiums.

Transport economists will need to look at the demographics both currently and in the future of the population including age, gender and other characteristics including family size. Economists will look at complementary services such as the Underground services to Heathrow as well as the Heathrow link railway and the possibility of linking Crossrail now renamed the Elizabeth line to the airport.

There are also bus services from Woking which make the journey easier from the national rail network. High Speed 2 could be linked to Heathrow.

Economists might also want to look at the taxation policy for aviation fuel and also for the other modes of transport. There could be a charging policy according to the volume of noise generated by aircraft.

They could be interested in the methods of mitigating the effects of aircraft noise for example Stansted Airport has bought a good deal of property around the airport so that there are fewer objections to airport expansion.

The location of airports to minimise noise would be thought out so that there is more flying over the sea or main estuaries rather than over major conurbations. This is one of the reasons why Boris Johnson when Mayor of London wished to have Heathrow and Gatwick replaced entirely by a new airport in the Medway estuary.

CHAPTER 13
Aviation

Size of Aircraft

The first commercial aircraft carried three or four passengers and with relatively low speeds of about 90 mph and were only slightly faster for cross channel journeys than train and boat. There has been a continuous increase in the size of aircraft. For example, in the 1930s, the average size was about 30–40 seats and this would be in marked contrast to the current European airbus, and the Boeing Dreamliner. Considerations of size are important for the government when considering the need for further airports since the fewer air movements to carry the same number of passengers the less will be the need for additional runways. Larger aircraft use proportionately less fuel, so the fuel shortage which seems likely to develop may mean that more attention is paid to this feature and less to that of speed.

Types of Aircraft

The piston engine airliner was used in the 1940s and 1950s but since that period jets have been used even for small executive aircraft. Bigger jets in the 1970s such as the Boeing 707 and DC-8 carried from about 150 to 250 passengers and had a range of about 6,000 miles with speeds of about 600 mph (just below the speed of sound). One of the major problems was that of noise. The then newer aircraft such as the Lockheed Tristar were considerably better in this respect and this will be important when considering the problem of location of airports.

The current so-called Jumbo jets, such as the Boeing 777, have a considerable range and are also fairly quiet and generally fly at much the same speed as earlier Boeings (i.e. just under 600 mph).

Noise of Aircraft

At the time of the Roskill Commission in the early 1970s people were reassured that in future aircraft would be so quiet that the noise from aircraft could reasonably be discounted. This was untrue. Information about the number of people who are likely to suffer from transport noise can be found for example in the current edition of Transport Trends.

Modern Aircraft

Some of the newer aircraft have only two engines compared with four engines in the past. The Boeing 777 planes have a range of 8,820 nautical miles. There are some restrictions on the routes which two engine aircraft can take compared with four engines, although the total power is about the same. This is partly because of the risk of engine failure so that the planes can be diverted if necessary to another airport rather than taking the direct route if it would not be near an airport. A four engine plane such as the European Airbus does not have this disadvantage. The Boeing 787 which made its test flights in 2007 carries about 210–310 passengers depending upon the way it is configured but perhaps most significantly claimed to consume about 20% less fuel than a comparable sized aircraft in the past.

Range vs. Payload

To some extent the air operator has a choice between range or payload (i.e. if they used more fuel, they will have a longer range but a lower payload). There are therefore many effects on the cost structure and the way that aircraft can be run. One of these is that unlike the railway system where overtaking is difficult with dual track unless there are passing places, the same is not true of the air system. If a plane starts out late from Heathrow Airport it may be able to catch up time if the airport at the other end is not very congested. On the other hand, since there are not the same defined national boundaries as with the rail or road system there has to be an agreement over how the safety of planes can be guaranteed when going from one country to another and therefore there has to be some method of providing navigational aids. This is one of the reasons why there has been more discussion about air transport at international level than about roads.

Concorde and Airbus Compared

Speed is the main advantage of air transport. Total capacity in terms of passenger miles or freight ton miles will depend partly on size and partly on the speed of aircraft (e.g. the Concorde when it was built in spite of a low passenger load was able to take as many passengers across the Atlantic as the large passenger liners because of its high speed – 1400 mph or almost Mach 2). Concorde was not a commercial success in spite of praise for its technology from parts of the media. At the present time, the costs of the European Airbus with its gigantic size have been rising rapidly and it remains to be seen whether or not it will be commercially viable. The cost of the European Airbus is spread across a number of countries.

Technical Spin-Off Insufficient

One of the advantages that had been claimed for the £1 billion plus program for the Concorde was that the research and development would be useful elsewhere. Both the Plowden report in the UK and the US's supersonic transport (SST) review committee have argued that this agreement cannot be put forward as a major justification for this project.

Supersonic transport in the past was represented by the Concorde, a joint Anglo/French venture in which the British Aircraft Corporation played a large part. It is noticeable that whereas Concorde was an Anglo French venture the European airbus costs are spread across a much wider variety of countries reflecting the high costs of R&D. This is in contrast with shipping where traditionally ships have not been at all standardised.

Trade-Off Between Speed and Size

There is a trade-off between speed and size, Concorde would have only just over 100 passengers but was able to fly at about Mach 2 about 1300–1400 mph.

The Problems of Supersonic Transport

One problem is the so-called superbang which can occur for 4 reasons:

1. Transition through Mach 1.
2. Atmospheric focusing so that about one house in 500 in a bang area might experience this.
3. Reflection effects, if for example it strikes smooth hard ground.
4. Manoeuvres by the aircraft such as changes of direction or changes in speed.

The effect of sonic bangs is important, because it may prevent Mach 2 aircraft from flying over other countries either losing the advantages of high speed or mileage to be added for example in December 1970 the US Senate voted to prohibit commercial supersonic over flights. Sweden, Norway, Netherlands, West Germany and Switzerland indicated at an OECD meeting in February 1970 that they were unlikely to allow over flying over their territories.

Limited Number of Airports Available in Emergency

Because they require long runways and specially strengthened surfaces, few airports could accept a Secure Sockets Layer (SSL). Presumably in these circumstances the conventional aircraft at larger airports would have to be diverted.

Light Planes

Whilst more attention has been paid to larger aircraft, smaller aircraft have often been used especially in the USA for executive flights (e.g. about 12,000 light aircraft are sold there each year).

The most well-known British light airplane was probably the Islander. The main advantage of such aircraft is that they reduce the accessibility time which would be required to go to larger airports as well as reducing the waiting time so that in many cases this will give the quickest possible journey time of any mode of transport. Southend Airport in Essex on the Thames Estuary has a considerable amount of light aircraft as it reduces cross channel times quite considerably in many cases.

Small Business Jets

There are several examples of small business jets (e.g. the Skyjet). Smaller aircraft are not without their problems, however. Some airline pilots have claimed that light aircraft were a menace to safety but this risk was ignored except when major crashes had occurred. In many cases, airport dues vary according to the type of aircraft.

Short Take Off and Lift Vertical Take Off and Lift (STOL and VTOL)

The main problem with VTOL has been to find a method of vertical take-off which can also be combined with the speed of fixed wing aircraft. There is little point in additional speed if access to and from the airports is a large part of total journey time so that in recent years more attention has been paid to the possibility of STOL or VTOL aircraft. One present limit on the maximum possible speed that aircraft can reach is that apart from exceptional circumstances such as spacecraft there is a limit to which the body can stand in the way of acceleration or deceleration without acute discomfort so that for this and other environmental reasons the SST seems unlikely to become the norm.

Seaplanes

These were often used since they provided a means of transport which did not need expensive terminals. Dornier currently produce some seaplanes.

They have generally been abandoned as a form of transport except where there are lakes or in the case of the Greek islands or the Caribbean where they can still provide an efficient form of transport. One of the problems has been how to get people to land afterwards. This is particularly acute if the seas are choppy. They are still used by rescue services where other forms of transport would not be suitable.

The Importance of Speed in Air Freight

Passenger transport has been predominant and freight has been subsidiary, unlike other modes of transport (e.g. shipping and rail). Traditionally most aircraft have been designed for passengers though the military still use freight only planes. There have also been quick-change aircraft which are aircraft which can be converted from passenger to freight.

In the past air freight has been mainly considered for the transport of perishable or high value products but in recent years more attention has focused on the concept of total distribution costs and the importance of reducing the costs of stocks, etc. This is possible where air transport gives both a frequent and fast service. If the service is not frequent, it will mean that in emergencies a sudden increase in demand for consumer products will mean that demand cannot be met.

The shipper may find that shipping in small lots is more expensive than in large quantities so some transport writers have suggested that customers could use sea transport for regular consignments and use air for non-regular consignments.

It has often been claimed that insurance costs are less by air than ship because the journey is quicker and therefore less susceptible to pilfering than on the longer ship journeys. This is not necessarily a natural advantage of air. This has been somewhat offset in recent years by the number of airport robberies.

Apart from perishable products, such as day-old chicks, flowers and fruit, other cargo such as animals (e.g. race horses or animals to and from the zoo), may require a quick flight because of the risk of damage. For fashion goods and where the weather or atmosphere could cause damage (e.g. the carriage of new cars or motorcycles), there will be advantages in a quick sheltered journey.

Using Aeroplanes as Part of the Total Journey

Sometimes, the advantage of speed may be combined with a low-cost form of transport such as coach to avoid a sea crossing which many people dislike (e.g. this may be the case from England to France, Belgium or Holland). However, the Channel Tunnel and the shuttle have competed vigorously with this market.

Sometimes as with the rail/air service using Southend Airport, speed of aircraft need not be very great for short journeys to give the fastest times for short continental journeys (e.g. to Belgium as the reduced accessibility time will more than compensate for slower flight speeds).

The International Air Traffic Association

The International Air Traffic Association (IATA) was formed in 1945 and its membership is composed of individual airlines.

The basic concept of IATA is non-political (i.e. the government may choose to have publicly owned airlines or privately-owned ones or to license certain airlines). IATA will try to ensure that there is a coordinated system of fares and charges.

The negotiation of these fares and charges took place through three area traffic conferences which generally take place at the same time because many airlines were involved in more than one area (e.g. British Airways will run flights to all three areas). The three areas were:

1. North and South America.
2. Europe, Africa and the Middle East.
3. Asia, Australasia, and the various Pacific Islands.

Another function of IATA has been the clearing house system whereby debts will be settled between the different airlines. Total turnover is very high partially due to the interavailability of tickets and freight being taken by more than one carrier.

During the early 1970s, there was increasing pressure from individual airlines who felt that fares on the North Atlantic Routes were too high. The problem was accentuated to some extent since chartered flights were not subject to the IATA recommended fares and therefore bogus organisations were sometimes set up so that potential passengers could take advantage of the cheaper fares.

Chartered flight operators were able to charge cheaper rates because by doing so they often managed to get better load factors which more than compensated for the lower fares.

Airships

These were used fairly extensively in the past and were "lighter than air" craft. They gained their lift from using either hydrogen or helium. Hydrogen had the disadvantage of being inflammable. They were much larger than the then existing aircraft (e.g. the British airship R. 101 was about 800 ft. long).

The typical speed of an airship in the 1930s was about 70–80 mph and they were used for trips across the Atlantic and competed to some extent on comfort with passenger liners.

They were abandoned because of their poor safety record. In recent years, there has been an increasing interest in their return.

Chartered Flights

This is where the flight is chartered by an organisation who will then sell the tickets separately to the members of such organisations. It can be where the aircraft is chartered by a travel firm.

Chartered flights have been used extensively for the so-called package holidays (i.e. where the customer pays an inclusive charge for travel and accommodation).

Currently in the UK, BMI has according to the latest statistics (2017) made available, a considerable impact upon this market.

The Problem of Chartered Flights

The increase in the number of chartered flights was due to the cheaper fares that could be charged compared with the ruling of IATA. There had been problems in that the British government had stopped some chartered flights because passengers were not always members of bona side organisations.

If this occurred under the then existing legislation the whole flight might be delayed. It was felt that the British government had enforced these rules more strictly than other governments (e.g. on one occasion people flew to the UK from the US without any query but were then delayed on their return journey).

Air and Rail Competition

On domestic routes within the UK the advent of electrification and the introduction of the HST have meant that it is difficult for the airlines to compete on intercity journeys of less than about 300 miles. The electrification of the lines between London and Manchester and Liverpool hit the then BEA services very hard. The West Coast rail modernisation process in the early 2000s was vastly over budget, but the results have been to make them more competitive in terms of time compared with the airlines. In Japan, the Tokaido line which covered the 300 miles from Tokyo – Osaka in 2.5 hours in the 1970s hit the airlines very heavily. Since then the expansion of the bullet train system as well as their punctuality has hit the domestic air travel market considerably. The frequency of the train system also makes this popular with many customers.

One of the reasons for the interest by the UK government in High Speed 2 is that it will help to keep to carbon emission targets.

IATA and Non-Price Competition

Because airlines were often restricted on the fares that they charged, they tended to compete on service rather than fares. This competition sometimes took the form of a choice of films or more lavish meals in the provision of newer types of aircraft.

This type of competition has been criticised on the grounds that if one airliner introduces a new type of aircraft other airlines will be forced to follow so that there is some degree of waste involved. The other main point is that some airlines have thought that the airfares have been too high and that this is one of the reasons for the lower load factors for many years. However, nowadays with the so-called budget carriers the emphasis has often gone the other way with complaints that in some cases wheelchair users have had to pay extra etc. Ryanair in particular has become well known for its no frills policy which on one occasion even went as far as no ice being available for drinks. In addition, Ryanair also charge £1 or €1 for passengers to use the toilet.

In September 2017, Ryanair cancelled many thousands of flights, including the Christmas period and the regulator issued a warning saying that they had not fulfilled their contracts with passengers.

Noise on Landing

The Department of Trade and Industry publication "Action Against Aircraft Noise" progress report 1973 stated that whilst many people consider that aircraft come in too low the need to follow radio beams which do not bend means that usually aircraft have to follow an angle of not more than 30°. There are still concerns about noise on landing and take-off.

Variety of Planes

Apart from the R&D costs limiting the number of types of aircraft, airlines will often wish to limit the types of planes they fly since the cost and time taken to train pilots will be a large proportion of total costs. There may also be problems with the provision of spare parts and training of mechanics if too many varieties are used. On the other hand, different sizes of runway may mean that some of the larger airplanes will not be able to use the airports so a variety may be necessary.

High Landing Speeds

In aircraft, generally the risk of accidents at landing increases proportionately to the speed of landing (i.e. landing at 100 mph will be 8 times as likely to lead to accidents as a landing at 50 mph).

International Civil Aviation Organisations (ICAO)

This body was set up after the 1944 Chicago conference and its membership consists of a large number of Individual states. It is now regarded as a specialised agency of the United Nations (in this regard it is comparable to IMO for sea transport). There are a number of committees which will report on legal problems, air navigation and other aspects of air transport. It will also investigate air accidents systematically.

The Five Freedoms

The Chicago conference which met during the Second World War in 1944 was concerned that international air transport should not be hampered by individual governments.

The five freedoms which were accepted were:

1. The right to fly over the air space of countries (who had signed the agreement) without stopping.
2. The right to make a technical landing i.e. for non-traffic purposes such as refuelling in any of the countries that had signed.
3. The right of country A to put down traffic in country B provided the traffic had been picked up in A.
4. The right of country A to pick up traffic in country B providing the destination was in country A.
5. The right of country A to carry traffic between countries B and C provided they had signed the agreement.

One restriction on these freedoms was that where the point of origin and destination are under the same government then only aircraft registered by the government can fly on that route. This would usually mean purely domestic routes though in some cases this could mean flying across another country's territory (e.g. before Bangladesh became independent, flights between East and West Pakistan would have flown over India).

Competition and the Airlines

One of the problems is that if the projected increase in air traffic does not occur (i.e. because of the problem of fuel prices) and climate change considerations were that there would probably be spare capacity (i.e. load factors would be low and that it would be easier to cope with this problem the fewer airlines involved).

Viability of Independent Airlines

Whereas at one stage many countries did not wish to have independent airlines competing with their own nationalised airlines, airlines and governments have altered their position so that in 2007 the open skies agreement between the European Union and the US was signed.

Easier Access for Passengers

For short journeys the faster speeds of aircraft have often been nullified by the time, which people have to report before departure of the aircraft. However, currently security concerns have made waiting times longer and it is uncertain whether Brexit will create even longer queues at both European and British airports.

More Standardisation of Fleets

One of the problems of many airlines is the great variety of planes that may exist in a single fleet (e.g. British Airways Aircraft ranged in 1974 from planes such as the Viscount first introduced in the early 1950s on European routes to the 747s). The advantages of standardisation would be to reduce stocks necessary for repairs and also the amount of training for maintenance.

Open Skies Agreement 2007

In March 2007, the open skies agreement took place. The idea behind this was that there should be fewer constraints on aircraft flying to and from the US whilst airlines such as the Association of European Airlines (AEA) welcomed this but there were some concerns about the environmental aspects.

CHAPTER 14

Safety

The Need for Systematic Evaluation of Accidents

Safety is an important feature of transport. However, the evaluation and prevention of accidents is rarely systematic. There is a bias towards looking at infrequent large-scale accidents such as the Kings Cross disaster on the London Underground in 1987, "Herald of Free Enterprise" the ship which was lost outside a Belgium port in 1987, but relatively little attention is paid to road accidents. Road accidents have fallen in the UK, but nevertheless there are about 1,500 fatalities currently.

This is much less than the typical figure of 7,000 fatal accidents and 350,000 injuries each year in the 1970s. This is partly due to better vehicles and partly due to safety measures such as the use of safety belts which has considerably reduced the number of deaths to car occupants. The UK figures are proportionately less than the majority of other European countries.

One of the major differences between attitudes towards accidents on different modes is accident investigation. On UK railways, the railway inspectorate will carry out an enquiry and will make suitable recommendations if any practices need to be changed. For aircraft crashes, which are comparatively rare, the so-called black box within the aircraft can often give valuable information about the likely cause of the accident. There have been few accidents on the Underground system though the fire in Kings Cross Station in late 1987 led to an extensive enquiry. Some press reports have put the cost of this enquiry at £4 million. In contrast, however, road accidents have very little investigation. It is up to the relevant chief constable as to what investigations if any take place. There have been suggestions, therefore, that there should be a more systematic valuation of accidents perhaps carried out by the Department for Transport (DfT).

Evaluation of Accident Rates

There are a number of ways of evaluating accident rates. One of the commonest but not necessarily the best is to look at the number of injuries or deaths per billion passenger kilometres travelled. On this basis air transport is undoubtedly the safest with less than 0.1 per billion passenger kilometres and motor cycling is the worst. Motorcyclists are roughly 38 times more likely to be killed in a road traffic accident than car occupants, per mile ridden. In 2013, 331 motorcyclists died and 4,866 were seriously injured in road collisions in Great Britain.

This analysis of accident rates may be sensible if we are looking at the best method to travel from A to B by different modes of transport. Economists are not necessarily comparing like with like, since many journeys that are now undertaken by air would probably not have been taken by other modes of public transport. This would include many holidays from the UK to the Caribbean and even more to Australasia. If transport economists assume that there are time constraints to the amount of time which people want to spend travelling to and from holidays, they could try to evaluate risk by the likely accident rate per unit of time e.g. one hours travelling. If as some people suggest we get space travel as one form of holiday, then this will become even more important. One accident per 1 billion km of space travel would be a high risk per hour, whereas if cycling it would be a very low risk.

Some modes of transport will improve health so that it is probable that if people are put off cycling because of fears of safety the improvements of health through cycling may more than offset the risk to health of accidents. The choice of words may be important. In an otherwise excellent report in the Green Paper on transport in 1977, it was stated that cycling was dangerous. This is untrue. What is dangerous is mixing cycling with other forms of transport.

How Safety Could Be Improved?

A systematic way of trying to improve safety would be to look at control of the way, control of drivers, pilots etc., control of the vehicle, control of the load, controls over managers, and investigation of accidents across the modes of transport. Transport economists could then see how safety could improve in a cost-effective manner including possibly transferring traffic from one mode to another mode or possibly within a mode of transport or even the timing of journeys since accidents occur disproportionately more at some times of day.

Control of the Roads

Another method of trying to reduce accidents would be to reduce the different types of traffic that can use the same highway. This is done with the motorways where pedestrians and cyclists are not allowed. This is in clear contrast to many roads in developing countries where there is a mix of all types of traffic.

Methods of Improving Road Safety

The majority of accidents occur at junctions. The roads can be made safer by minimising right hand turns in the UK as they go across the flow of traffic and in some cases to have grade separated crossings. This means vehicles do not cross each other's paths at the same height. This would be left hand turns in the majority of other countries where driving is on the right. Another method is to restrict the number of intersections as is done with the motorways, which is why motorways have a good safety record. It is difficult at the moment to insist on suitable headways between vehicles though the Highway Code in the UK gives advice for this which is often not heeded. Speeding is a wide spread occurrence and leads to the risk of serious accidents which is increased significantly with higher speeds. This was shown very clearly after the oil price rises in 1973/74 when for fuel economy reasons lower speed limits were introduced in many countries including the US and as a result the number of accidents fell.

It is possible that in the future that computer controls would indicate to the driver perhaps audibly that there is insufficient distance between vehicles. There could be a modification of satellite navigation to do this more extensively. At the present time satellite navigation can give distances of vehicles within about 15 metres. A newer system proposed by the European Union (EU) would be able to pinpoint a vehicle more precisely. The EU wishes to have a Galileo positioning system more often called simply Galileo to have a global navigation satellite system. The advantage of such a system compared with the ubiquitous satnav (referring to satellite navigation system), is that because it would be more accurate than the existing system, then it could give more information in really crowded areas. There have been points highlighted by the media, where satellite navigation has led to people being over reliant on it, e.g. going down a railway track rather than by road, and in some cases going on totally unsuitable roads because the satellite navigation system does not take account of this.

If Galileo comes to fruition, then this might make it more reasonable to use it in the future with more precision nearer intersections whereas at the moment voice control could be used on open roads rather than at junctions where precision is more necessary.

The reduction in the number of intersections, adequate visibility, and the avoidance of hump backs may also help to minimise accidents. However, the majority of accidents are not the result of the way being inadequate.

Computer controlled cars and lorries could be safer.

Driver Error

Driver error is still the main cause of road accidents. Whereas there are strict controls over public service vehicle and heavy goods vehicle drivers, this is not true of ordinary car drivers. There have been in recent years in the UK a tightening up of the driving test for motor cyclists. The test for car drivers has been extended to include a multiple-choice element as well as driving on the roads. This was extended in September 2007. The Highway Code issued in 2015 mentions the need for concentration including when smoking. There have been suggestions that rather than having a driving test which gives a lifelong licence that there could be retests. This would be resisted fiercely but there might be more scope to have a system of eye tests which would be fairly easy to implement. The logic of this is that eyesight deteriorates with age. Whilst people are asked whether they can read number plates at suitable distances during the driving test, there has been little evaluation to show how far eyesight deterioration contributes to the number of accidents. Most people might assume perhaps correctly that if people have poorer eyesight that they are going to find it more difficult to be aware of what is going on, and therefore this could be dangerous. However, there has been little academic interest in this. If, as one might expect, eyesight does deteriorate with age, which would seem to be plausible given the number of older people wearing spectacles then one would need to have a cost benefit analysis of seeing how far and how much it would cost to have re-tests compared with the likely savings in accidents. Texting is a new and dangerous form of distraction for drivers (whether they are driving lorries or cars). In 2017, about 200 drivers a day have been stopped for using mobile phones whilst driving.

Transport Select Committee Report

In July 2007, the Transport Select Committee suggested that the age for a driving licence should be 18, that new drivers should not be allowed to have young people as passengers and that there should be a zero limitation on alcohol for the first year. This was because though the 18–25 driving population is a small part of the total number of drivers there are over 1,000 fatal accidents per year. The House of Commons Transport Committee conducted an inquiry into novice drivers, with oral evidence sessions being held in early 2007.

The chance of a driver being involved in a collision is particularly high during the first year after passing the driving test. The government estimates that nearly 38,800 people are killed or injured each year in collisions involving at least one driver with less than two years' post-test experience. The inquiry considered:

The Nature of the Problem

To what extent novice drivers are more at risk of being involved in a collision than other drivers, and whether this is primarily a consequence of age, inexperience or a combination of both?

Do young people's attitudes to driving have a significant impact on the collision rates of young and novice drivers?

Driver Education and Testing

How effective are the existing practical and theory driving tests at identifying safe driving skills and behaviour? Has the hazard perception test achieved its objectives?

Could changes to driver education and testing help to make novice drivers safer? Such changes might include: new pre-test requirements, such as a minimum number of hours' or miles' driving, or a minimum period between obtaining a provisional license and taking the test; compulsory professional tuition; or additional training for motorway driving or night driving.

Graduated Licensing

Graduated licensing schemes involve the phasing-in of driving privileges. Typically, a graduated licensing scheme imposes additional restrictions on new drivers either for a fixed period of time after passing their test or until a second test is passed. Restrictions in a graduated licensing programme might include: a lower speed limit; a lower blood-alcohol limit; restrictions on the number of passengers who may be carried; and restrictions on night driving.

Changes to the Driving Age

Would there be any benefit in changing the minimum age at which a provisional or full license may be obtained?

Different Treatment of Offenders

Drivers face disqualification and re-testing if they acquire six penalty points during the first two years after taking their test. Could further, similar provisions for the different treatment of novice drivers who offend be introduced?

Whereas nearly all the other modes of transport have stringent regulations about drinking and driving, relatively little has been done in the UK or elsewhere about ordinary car drivers. The estimates are that in the UK that about 30% of road deaths were attributable to drink and drive. In the UK, there were strong hints from the then government in 2007, that the drink limits would be reduced from 80 mg/100 millilitres of blood to 50 mg/100 millilitres (i.e. that even half a pint of beer would be the limit).

Both legal and illegal drugs may affect driving performance. Some countries however notably Sweden have very stringent regulations about drinking and driving. Whereas public service vehicles and road haulage drivers have hours' limits imposed upon them by EEC Regulations 3820/85 and subsequent legislation there is nothing to stop the ordinary motorist driving for long periods. Many drivers cannot concentrate for very long periods. The importance of adequate driving regulations can hardly be overstated. The 1976 Consultative Document stated that whereas vehicle failures were responsible for about 8% of accidents and deficiencies of roads for 28%, human error contributed to 95%. High speeds contribute to accidents and this is perhaps shown by the speed limits, which were imposed at 50 mph after December 1973 which reduced road accidents in comparable months very considerably. This decline was considerably above the proportional decline in traffic. Whilst it is sometimes claimed this would increase driving time considerably, research from the Stanford Research Institute in the USA indicated that the average time for journeys with a 20% reduction in speed limits would have been 11 minutes and would have reduced deaths in the USA by 40%, which would then have been 26,000 lives.

There was considerable pressure on the then Department for Transport to allow random breath tests. Many commentators feel that whilst this would be to a limited extent an infringement of personal liberty the deaths which would be prevented would outweigh this.

Motor cyclists have a very high accident rate. Part of the reason for this may be not only the lack of segregation for motor cyclists but also the age of the riders. A large number of deaths occur to riders in the 17–29 age group. There are also large numbers of cyclists' deaths perhaps partly for the same reason. There is a strong case for improving the safety training of children. This is already done to some extent through the cycle proficiency test. There have been arguments about the provision of safety helmets for cyclists.

The number of people killed on the roads has declined since the introduction of compulsory use of safety belts in 1983. Since then there has been controversy about whether there should be compulsory use of rear seat belts. Seat belts are now compulsory for people of all ages in cars and this has reduced the number of serious accidents and fatalities. Seat belts have been compulsory in almost all West European countries. However, many people still do not obey the regulations.

Accidents and Pedestrians

Whilst pedestrians do not cause the majority of accidents, they are disproportionately likely to get injured or killed. Those particularly at risk include the young and the elderly. Greater training for young people would be desirable. There has been considerable publicity about children's risks in going with strangers, yet the risk of road accidents is far greater.

Reduction in Accidents per Passenger Kilometres

The construction and use regulations have improved road safety and braking systems have helped.

For cars over three years old there is an annual DfT test (called MOT test) though critics have suggested that this is not rigorous enough. There are also tests for public service vehicles and road haulage vehicles.

Road accidents both in the UK and elsewhere, have been the cause of a great many deaths and serious injuries but except at Christmas when there is widespread publicity and sometimes at other bank holiday weekends accidents are often taken for granted. The number of road accidents has declined in recent years, partly because of traffic engineering (e.g. the use of traffic lights, junctions, bypasses) though driver behaviour counts for a large part of the present number of accidents. However, adequate separation between different types of traffic – pedestrians, cyclists, motorists and lorries – will reduce the number of accidents. This can be seen in Holland where there is adequate provision of separate cycle ways. Within the UK the provision of motorways, which eliminates right hand turns reduces the number of intersections and pedestrians, cyclists and car learners, is one of the reasons why in spite of greater speeds the accident rate per vehicle kilometre is lower than that for conventional roads. This does not necessarily, however, mean that this is the most cost-effective way of improving safety. It may be possible to reduce the number of intersections without the high cost of motorway construction. The former GLC, in its paper 'Traffic and Environment' showed that on two comparable roads, the A 10 and the A 1010, there were far fewer accidents where there were fewer intersections. If these were typical then it would indicate that on other main or arterial roads there should be fewer intersections even though this might add to the costs of transport. One of the effects of the lack of segregation has been frustrated demand. Children very rarely cycle to school even where it might be their wish because of perceptions of safety risks. The bubble cars of the late fifties and early 1960s and the introduction of the C5 pioneered by Sir Clive Sinclair in the late 1980s failed partly because of this vulnerability in mixed traffic.

Underpasses and Footbridges

The provision of underpasses or footbridges for pedestrians is sometimes made on a piece-meal basis. They can, however, be provided as a completely separate system of usage, as for example The Barbican Development in the centre of London. Thamesmead in South East London has a system whereby the main roads in the town do not have any houses directly going on to them. It also has a system of cycle paths and footpaths, which are quite separate from the majority of vehicle traffic. The concept of having pedestrians separate from vehicle traffic was suggested in the Buchanan Report (the more formal title was "Traffic in towns" 1963). The Buchanan Report's ideas were not implemented, however, because of the high cost. One estimate at the time was that it would have cost about £18 billion.

There may be however more cost-effective ways of reducing the number of traffic accidents than altering land use. The introduction of compulsory use of front seat belts was introduced in Great Britain in 1983.

However, traffic engineering measures such as computerised traffic controls where a large number of traffic signals are linked to a central computer not only helped to increase total capacity by about 10 per cent but might also help in reducing the number of accidents. This is because the traffic lights being linked will enable motorists going in one direction to have green lights all the time providing their speeds are not excessive.

The use of one-way roads may also be helpful since they reduce the likelihood of head on collisions. Head-on collisions are the most dangerous ones. They may also help to speed up traffic flows though they tend to increase the travelling distance. There are also beneficial side effects, noise levels may be lower and fuel consumption reduced since there is likely to be less starting and stopping.

Contrast with Rail Transport

The railways, particularly in the UK, have had stringent regulations about drivers and in the UK, drivers will have to learn the routes before they can go out in a train on their own which is usually called route knowledge. Until recently in the UK there has usually been a second driver which has partly been a hangover from the steam engine period when there was a fireman, as well as a driver. The railways have usually had a guard whose role has changed to being a conductor guard mainly in order to increase revenue and reduce costs to the railways.

Rail Safety

Transport economists will observe that the railways have generally a low number of accidents partly because the way and the drivers have usually been subject to overall control by one body although this has changed since privatisation. Professor Andrew Evans has suggested in an article in the Royal Statistical Society journey that safety has improved since 1997 if we look at the long-term trends, although Chris Wolmar, a well-known public transport commentator, disputes this.

It is easy for complex signalling systems to be installed. The headway between the trains is also generally sufficient to avoid collisions though some still occur as in Paris in June 1988 with the four-colour light signalling system of red, amber, double amber and green which is common on railways in the UK. There will be a minimum distance between trains although there have been concerns about signals passed at danger, although the extent of this seems to have fallen. They are subject to the same overall control, in the USA where Amtrak is responsible for the track on which several railways may run. This has become even truer since the privatisation of the railways in the UK. The Hatfield disaster (October 2000) when four people were killed is one of the problems which the rail industry has to face and some commentators have stated that using subcontractors is unhelpful. Network rail has often used direct labour rather than subcontractors since then.

In March 2007, Network Rail was fined £4 million for the lack of safety procedures which led to the Ladbroke Grove (near Paddington) crash in October 1999. This was criticised by some of the commentators including ASLEF which stated that as network rail was a not-for-profit organisation that the fine was paid by taxpayers and it would be better if managers had to take direct responsibility.

Because the number of rail deaths is very small, there seems little evidence that any particular locomotives or type of rolling stock are inherently more dangerous than others. However, there has been some concern expressed noticeably by the National Union of Rail, Maritime and Transport Workers about over-crowding on commuter services. This in itself is not necessarily dangerous but problems could arise with evacuation if such vehicles were involved with a crash.

Shipping and Safety

For shipping, much of the safety provision such as navigational aids and charts is provided by international or national bodies. However, there are problems in very busy international waters particularly through the English Channel.

The Torrey Canyon (disaster of 1967) showed that the costs resulting from a collision were considerable and would not necessarily be borne by the transport suppliers. The International Maritime Organisation has had a variety of load line conventions to try to prevent overloading of ships. This was originally known as the Plimsoll line, in the nineteenth century. One of the problems with larger ships has been that navigational charts have not necessarily been adequate because of the larger draught of the newer ships. The International Maritime Organisation has therefore co-operated with the International Hydrographic Organisation in providing adequate knowledge through better charts.

Air Safety

In aviation, air traffic control is of vital importance and because of the high speeds there is a need for an international language. English is used for this purpose. There are problems of ensuring that the air traffic controllers have adequate knowledge and in determining who is responsible for air safety in open skies, which is not within any one nation's territories.

Air and Sea Transport Compared

For air transport, the regulations about pilots and which aircraft they are allowed to fly have also been stringent. For shipping, there are a whole series of certificates of competency for different merchant navy ranks. There was some concern particularly in 1988 by the then National Union of Seamen about the adequacy of manning of passenger and other ships.

Statistics show that the safety record of flags of convenience countries has not been as high as that of non-flag of convenience countries. The International Maritime Organisation has tried to improve the quality of shipping staff through assisting the training of crews in under developed countries.

For shipping, standards of safety vary between countries. One of the problems of some flags of convenience countries is some would seem to have no adequate facilities for ensuring that ships on their register do comply with safety standards. For aircraft, the International Civil Aviation Organisation sets some standards. The risks for aircraft generally occur with landing and take-off rather than in mid-flight. It might be possible to reduce the severity of accidents through changing the type of fuel used. With aircraft accidents, smoke is often the cause of death rather than any fires themselves. There was considerable controversy during 1988 about possible methods of dealing with this problem. Some transport economists commented that whilst duty free was carried in most international flights there was concern about the greater weight of fuel which was less likely to cause these deaths and injuries.

CHAPTER 15

Coordination of Transport

The Aim of Economic Coordination

The aim of economic co-ordination is to meet the demands for a given level of transport at the minimum resource cost. Problems arise when trying to measure resource costs, since they are not always the same as accounting costs, either because there is a divergence between social and private costs or because there is a divergence between historic and current costs. The divergence between perceived and actual costs means that competition does not always lead to an optimal use of resources.

The nonprofessional often refers to coordination, such as the need for inter-changes, common timetables, the co-ordination of bus and train times and the suitability of containers for different types of transport. Economic coordination could occur through pricing policies or administrative mechanisms or a mixture of the two. The late Gilbert Ponsonby who died in 2012 argued that co-ordination could be achieved through competition. The Pareto optimum will occur subject to very stringent conditions. However, few sectors of the transport industry conform to the model of perfect competition. In particular, since no country as a whole apart from Singapore has a national system of road pricing which reflects resource costs, the road haulage and road passenger industry will not reflect these costs. London now has a road pricing policy, but this is only for a small part of the London area and a very small part of the UK landmass. Economic coordination whether through a competitive or an administrative system requires good information. When forecasting need for transport infrastructure, it is difficult to estimate potential demand. Rail future, a transport pressure group has highlighted in September 2016 newsletter how Lea Bridge station, Cambridge Heath and Hackney interchange London Fields have exceeded forecast demand.

One of the latest reports on High Speed 2 is 5200 pages long, illustrating the problems of forecasting demand. If demand is over estimated, excess capacity could remain for a long period. This is one of the reasons why there was at one stage in the UK a National Ports Council. Once ports are built they are expensive to eliminate and often derelict ports will have an adverse effect on the neighbourhood which can be seen in many areas. This is what the economist would call a negative multiplier effect. Excess capacity can arise for several reasons, for example, charges in demand following the introduction of newer modes of transport. Once the railways gained great strength, many of the canals which had been built during the canal mania of the late eighteenth century were no longer required. It may also occur when it is not technically possible to adjust capacity to meet demand (e.g. when new techniques such as improved signalling or containerisation necessarily increase capacity). The indivisibility of much investment means that this is almost inevitable.

Spare Capacity

Spare capacity may also arise when there are too many new entrants into the industry, as for example road haulage in the UK in the 1920s. Many of the people coming out of the armed forces in the First World War had gained driving experience as well as maintenance skills and assumed that they could use this in the road haulage industry or in the bus industry. This led to excess capacity in both these sectors. Spare capacity may arise if fares or charges are fixed at too high a level, the organisation IATA has been in the past accused of causing excess capacity. There seems no reason to assume that the equilibrium position, which is stressed in many conventional economic textbooks, will occur. This is not necessarily an argument against competition however, since regulatory bodies, whether they be traffic commissioners, local or central government will not necessarily have better information on which to base decisions. One of the weaknesses of the A, B and C licensing system which existed in road haulage in the United Kingdom prior to 1968 and road passenger transport licensing system prior to the 1980 and 1985 Transport Acts, is that whilst it may be easy to predict demand for a few months it is difficult to do so for longer periods.

Longevity of Transport Infrastructure

One of the problems of transport coordination is the longevity of transport infrastructure. Assets such as docks, railway and canal beds are very specific and cannot usually easily be used for other purposes. The Channel Tunnel is a good example where it is difficult to think of any use for the project apart from rail transport. Overprovision of these assets can lead to waste as has occurred with railways and canals, but under provision such as the failure to provide enough airport space could also lead to wastage. Often the vehicles themselves have a long life, and whilst these are usually less specific assets than the infra-structure, wastage can still be taking place. British Railways were building steam locomotives up to 1960. These had a technical life of perhaps 30/40 years, but virtually all steam locomotives were scrapped or removed from British Railways by 1968.

Knowledge of Costs and Coordination

Economic co-ordination implies that costs are known to the operator and the consumer and that prices reflect these costs. In many cases, this cannot apply as taxes or subsidies are imposed. There is also the particular problem of allocating costs for railways where there are a large proportion of joint costs.

The problem of perceived costs arises in the own account sector of road haulage and also for cars in road passenger transport. If owners underestimate their real costs, there seems to be no reason why competition will automatically lead to economic co-ordination and to the optimal use of resources.

The European Union and Marginal Social Cost Pricing

Ideally, a pricing system, whether for rail, road or air, would reflect social costs and benefits, including noise, pollution, congestion and accident costs.

This could be done if as suggested by the EEC (later the EU), there was system throughout all modes so that prices reflected marginal social cost pricing. The divergence between private and social costs does not necessarily indicate that public ownership is preferable. There seems to be no reason to assume that the National Freight Corporation, set up in the UK in 1968, had any greater sense of social costs than many private organisations. There is no reason to assume either that economic coordination will necessarily take place within an organisation which covers more than one mode of transport. There was little coordination between the Underground services and the bus services in London, though both were under the control of London Transport. Passenger Transport Executives on the other hand seem to have made some attempt to evaluate the relative merits of bus and rail services and have allocated resources accordingly. Lack of coordination does not apply to the public sector alone. The Railways act 2005 stated that they could specify service improvements as long as they also stated how these could be funded. They can also lobby for different improvements to different services. Own account transport firms' distribution including transport, has sometimes been regarded as a "Cinderella" function and there has been little co-ordination, if at all, between one part of a firm and another.

The late management expert Peter Drucker on one occasion commented that distribution was the last frontier in management.

Need for International Economic Coordination

Economic co-ordination should not be thought of solely in national terms. If there is a genuine fuel shortage, international coordination could make the best use of resources. This could include international research into the possibilities of greater fuel efficiency and possibly into alternative sources of motive power.

For international air transport, whilst there are objections to cartels, there seems to be little logic in having low load factors on aircraft if the same volume of traffic could be carried in fewer aircraft.

In order to achieve effective coordination, one would need to take into account cross elasticity within and between modes of transport when planning to invest in vehicles or infrastructure.

When looking at investment, economists need to be sure that even if investment seems to be worthwhile on a discounted cash flow technique, losses occurring in other parts of the transport system would mean that it was not worthwhile from the community viewpoint. For example, investment in say British Airways shuttle services, could by conventional investment appraisal methods, give us a higher rate of return than that suggested by the 1978 White Paper on Nationalised Industries. If, however, improved air services took a great deal of traffic away from the railways, then the UK economy may not benefit fully from such investment. Economists have to consider whether they are looking at the rate of return on historic cost, which in inflationary periods would give fairly high rates of return or in terms of current cost accounting when smaller returns would be shown.

Investment in the different modes of transport have been made on different principles in the UK. For example, the London to Manchester route, showed cross subsidisation by the then nationalised British Airways from the international to the domestic routes, whilst British Rail would have justified electrification on broadly commercial grounds. Investments in the motorway network would have been justified by cost benefit analysis. There is no reason to assume that using different criteria will lead to optimal results. Until recently, rule of thumb criteria has been applied to road investment in the UK, though in more recent years, cost benefit analysis has been applied to all road projects over £250,000. In spite of large-scale government intervention there has been little overall transport policy. Governments have looked at individual projects on too narrow a basis. For example, when looking at the third London airport site, alternatives such as greater use of regional airports or extensions to the existing London airports were not considered. Other developments, such as the Channel Tunnel, which affected the need, were disregarded. Except for road investment, private rather than social costs have been considered.

Coordination applies to some common form of pricing criteria. It is not obvious whether this would mean short run marginal cost pricing which would make the best use of existing capacity, or long run marginal cost pricing which gives better criteria for investment.

Too rarely have governments looked at the most cost-effective way of providing transport. Often the emphasis has been on the supply side of transport, stressing the need for mobility, whereas it might have been cheaper to have provided accessibility. There has often been no cost benefit analysis when considering the effects of larger hospitals and schools. Sometimes higher transport costs may have outweighed the savings to health or education authorities. Transport is also affected by the political ideology of the government of the day as the Labour Party has been more willing to support public transport while the Conservative Party traditionally supports the private transport industry.

Economists may look at many company cars. This subsidy has been largely overlooked by government and many transport commentators, whereas subsidies to public transport had much greater scrutiny. The latest copy of Transport Trends however suggests that only about 9% of the total fleet can now be classified as company cars.

How Can Costs Be Effectively Allocated?

In several modes of transport, costing have been poor, often because of administrative difficulties. To use the railways as an example, it is difficult to decide how costs should be allocated to trains, consignments, passengers, or complete services. There has also been a failure to look at avoidable costs when considering closure. Whilst for example railways are carrying traffic there are problems in allocating overheads such as the cost of administrative staff, especially where railway lines carry both freight and passengers. When considering closures, one needs to be aware what resources are actually saved rather than merely looking at conventional accounting costs. The then Independent Transport Commission in their book published by Coronet Press, 1974 stated, that Licensing systems restrict entry mainly to existing entrants and also have side effects, for example, firms may buy out other firms in order to obtain licenses. A quality licensing system on the other hand, as with the 'O' license system which has applied in the UK since 1968, has a much greater emphasis on safety. The licensing system for road passenger led to cross subsidisation. Sometimes the licensing system could insist on physical coordination though this is difficult if there are different size units of carriage. The 1968 Transport Act had clauses whereby lorries over 11 tons and travelling more than 100 miles could have been asked by the appropriate licensing authority whether the trip could have been carried out more efficiently by British Rail or the National Freight Corporation. These clauses were not implemented and were repealed by the 1980 Transport Act.

There have from time to time been suggestions that licensing authorities may assist if greater scope is given to transhipment depots and groupage, etc.

Coordination of Research

To some extent research might be thought of as a collective good and duplication of research can therefore be wasteful. In this case, research such as that undertaken by the Transport and Road Research Laboratory may lead to better utilisation of resources than if it were carried out by a series of separate firms. Coordination of international research across the European Union and beyond is important.

CHAPTER 16
The Public Sector

Definition of the Public Sector

The public sector is the part of the economy which either directly or indirectly is run by central or local government. There are a variety of different types of enterprise within the public sector dealing with transport.

Public Corporations

Nationalised industries are also often referred to as public corporations and have played a large part in the transport sector. The organisation of the public corporations is laid down by Parliament in the relevant acts, for example, the 1947 Transport Act was responsible for the setting up of the British Transport Commission (BTC). The Acts of Parliament usually specify that ministers may appoint the chair person and members of the board whilst the day to day running decisions will be taken by the board and other members of the public corporation. The Acts of Parliament may also specify economic objectives, the nationalised industries including the transport sector were expected originally to break even. However, even if an organisation is expected to break even, it is not clear whether profitable services should subsidise unprofitable ones, usually called cross subsidisation or whether individual services are meant to break even. If there is cross subsidisation many economists would ask why people on the main lines should subsidise people on other routes. Sometimes there may be cross subsidisation between different times of journeys. The assumption was that the peak was profitable. However, with greater knowledge of costing, it was clear that often the journeys between say 10:00am and 4:00pm particularly on the buses were profitable and these cross subsidised the peak times. It was even less clear why the people in the peak (often the more affluent) should be subsidised by other travellers. Sometimes, it would have been difficult anyway to have made profits on the more sparsely used services, so this interpretation was unlikely. In recent years, White Papers or Acts of Parliament have specified the returns to be made on new investment. White Papers or Acts of Parliament may also specify a range of activities which can take place within the corporation. One of the criticisms of nationalisation was that often it led to a more rigid organisation than would have occurred on commercial criteria. With the decline of the railway services monopoly as such in the inter-war years with increasing competition from the road freight and road passenger sectors the private railway companies reacted by buying bus and road haulage companies until by the time of nationalisation the railways between them were some of the largest road passenger and freight organisations. In the last few years, there has been a considerable debate about nationalisation and privatisation. The nationalised industries were set up originally for a variety of different reasons. These include:

Control of Monopoly Power

Where it was felt that there might be a degree of monopoly power which might be undesirable in private hands. This applied to other public corporations such as electricity but was irrelevant to most of the transport industry where British Rail had lost most of its monopoly power by 1962.

Better Industrial Relations

It was hoped that where industrial relations have been poor in the past that working for organisations which were not geared solely to profit might improve them. Some economists might suggest that it would have been better to syndicalism (i.e. where workers controlled the organisations rather than the form of nationalisation which Herbert Morrison, the then Labour Minister, had decided upon).

Strategic Reasons

In both the First and Second World Wars, the railways were used extensively to carry both troops and equipment as well as to carry normal passengers and freight. It could be argued that subsidies to air transport such as that to Concorde seemed to be more geared to looking at defence needs rather than to the needs of passengers.

Targets and Effects

The logic of breakeven is that it would have given the advantage of economies of scale without the disadvantages that private monopoly would often have restricted supply in order to boost prices.

The railways were nationalised in the 1940s partly because they had been under government control during the Second World War and had been intensively utilised. During that period, there were insufficient resources to devote to improved trains and rolling stock so they were in a rundown condition.

British Rail Freight Services

One of the developments on British Rail was the merry-go-round trains whereby 200 hopper wagons did the work previously done by 2,000 conventional wagons. By far the majority of coal which is transported to power stations was transported in such trains. The advantage of the hopper wagons was that they did not have to stop to load or unload, as they can go at very slow speeds whilst these processes are being carried out. The coal-miners' strike during 1984–1985 reduced the total market for coal. The privatisation of the electricity industry in 1988 also reduced the demand for coal generally and hence the capacity of rail freight. In practice, the rundown of the coal industry has been very fast and so the main coal industry traffic now carried is to and from the ports.

Company Trains

These often operated on a "sidings to sidings" basis such as between Fords in Dagenham and Halewood. Using company trains means that the railways can carry large quantities at relatively high speeds generally over long distances. Company trains have been used for the carriage of cars, steel, chemicals and some mineral traffic. The railways also avoided the costly system of having many marshalling yards which had existed until the Beeching plan 1963 after which many of them were closed down. Marshalling yards are where wagons could be shunted from one train to another. There were too many of these. Carlisle had 9 marshalling yards which was excessive even though it served many routes. The advantage of direct services means that since the trains are used effectively with a large load, even fairly short distances can be more competitive by rail. This is in contrast to the generally stated idea that the rail services can only compete for distances of around 150–200 miles.

Parcel Trains

British Railways sometimes used special trains for the carriage of parcels. Sometimes however the traffic has been conveyed on passenger trains sometimes in the brake compartment or sometimes with special stock which had been added to the passenger carriages. For a long while the Post Office had its own trains whereby the post could be sorted on route so that the overall time for the post could be improved. It was felt that the withdrawal of the Post Office trains showed the need for a cost benefit analysis approach since clearly more road traffic would add to social costs. In 2008, Netherlands was experimenting with the use of freight only trams which could be used particularly in the evening and night to transfer the smaller items of freight.

The Automation of Railways

In the 1970s, British Railways originally hoped to have an auto-wagon system which would have avoided the need for drivers on some freight trains. This would have had significant advantages when carrying less than a complete train load of traffic. However, this was not pursued. Having unmanned trains travelling at low speeds using the many private sidings could have reduced costs of transhipment and helped the railways gain more traffic. It is more possible to use unmanned trains where systems are self-contained as for example with the Post Office which had its own Underground railway in Central London for 80 years. The Docklands Light Railway is a good example of a driverless train, although it does have a train controller.

Speed Link Surveys

Not all rail freight can be carried in train loads. One possibility would be therefore for the rail operators to forego this type of traffic altogether. This, however, would mean that it would lose large amounts of freight. British Railways therefore developed a system called Speed Link which used higher capacity wagons than in the past. The speeds were also quite fast and were even faster once the electrification of the East Coast main line had been completed.

The total operations processing system (TOPS) computerised monitoring system also meant that it was much easier to check where individual wagons are. This avoids the embarrassment which was sometimes caused in the 1960s and earlier years where wagons could be reported lost. Ideally Speed Link is used when both the originating point and the destination have private sidings since this will give the minimum of handling and would also enable a service to compete more readily with road freight on a total time basis. The Speed Link services tried to minimise the use of marshalling yards.

Computerisation of Rail Freight and Rail Passenger Information

The TOPS system of computerisation stands for total operations processing system and enables the railway to monitor its freight traffic. The system was originally devised and developed in the USA. The use of freight concentration depots rather than from individual stations means that there are less freight centres around the UK to feed information into the computer. The information can be read directly from the wagons to the central computer system.

On French railways seat reservations are compulsory on TGV and some other trains. The seat reservations can be made through the appropriate offices at many stations or can be made by the passenger himself at the station. The passenger in this case would have to show his or her requirements by pushing the relevant buttons on the machine at the station.

In the UK, a great many passenger stations are now linked so that it is possible for passengers to reserve seats on trains and to obtain the relevant seat reservation very quickly. The advantage of computers is that the information can be passed to the operator so that it is easy to see where the trains are fully laden and where there is spare capacity. Recently, South West trains have offered some very cheap journeys between Portsmouth and London, presumably because they have identified some services where there is still slack. These have to be ordered via the web. The advantage of this from the railways point of view is that it gives much greater flexibility in pricing than has been possible until recently. The railways wish to use smart cards which could not only be used for rail and other public transport journeys but could well be used for many other purchases, rather like loyalty cards which are used by the supermarkets computer data can be used to analyse total journeys made.

There are also opportunities to be able to sell information to third parties which may be a useful form of revenue.

Computer data is also used with announcements at stations so that it is easier to identify if there is late running. This may be very useful to passengers especially if they have a choice of routes so that if there are delays on one line they may be able to use another line or route.

Other Methods of Trying to Reduce Railways Costs

The railways have tried to reduce the maintenance costs of the system by using continuous welded rail rather than the traditional 60 ft. lengths. They have further developed the concept of paved concrete track (PACT). These trains can go along the track and detect where there are any weaknesses. The railways have also used preventive maintenance in order to reduce costs.

Use of Modern Technology Elsewhere

Some of the rail operators are also using mobile phones so that they can be used to get tickets which has the advantage for the user of not having to wait at the booking office. Currently, it has been estimated that more than half the world's population own a mobile phone. Therefore, the use of mobile phones for ticket purchase is likely to become more common.

The Use of the Internet

This can be helpful with the public sector as with other transport organisations so that people sitting in their own houses can work out the best possible routes. With many mobiles having Wireless Access Protocol (WAP) so they can access the internet, people can also do this on the move.

The government has tried to have an overall transport website so that people can see the best methods of travelling whether they are using rail, Underground, buses or even in some cases ferries. However, there have been some teething problems. In many cases, nowadays people cannot merely see the costs of the journey which they want but also see if there are cheaper prices if they go at slightly different times. This is helpful since on many journeys there are a wide variety of fares. For example, currently by rail from Kent to Bristol, there are at least 14 different fares.

Privatisation of Transport Systems in the UK

One of the features of the 1990s was privatisation. It was assumed that this would tend to reduce the public sector-borrowing requirement (PSBR). This became more important after the Maastricht treaty 1992 within the EU required countries to limit their PSBR as a proportion of national income. Some people might however suggest that this is somewhat artificial. For example, with the Private Finance Initiatives (PFIs) that some of the money which comes from the public sector in any case is no longer counted as public sector since it is channelled through the public companies. It was also assumed that privatisation would lead to growth in shares and this was an objective, which particularly the Thatcher government (1979–1990) had as one of its aims. It was sometimes assumed as well that employees might want to hold shares, and this was certainly the case with the National Freight Company later called the National Freight Corporation. However, it is difficult to be certain about how far share ownership was encouraged in the long term as opposed to people who brought shares in the hope of quick capital gains. In this case, we could test whether the objective of share ownership had been achieved by looking to see whether people had still retained shares perhaps five or ten years later. If they had not done this then clearly the objective had not been achieved. The subsidies to the railways, in the United Kingdom are currently higher in real terms than when British Rail was a nationalised industry.

The Social Service Arguments

One of the problems of nationalisation had been that the government had often intervened irrespective of the legislation. For example, in some cases there were social aspects such as the provision of rural rail services. Whilst this might have been desirable it would have been better in many cases if there had been transparency of subsidy (i.e. the public could see what it was paying for and why). The subsidies to the railways are much higher than before the railways were privatised and it is not obvious that the subsidies necessarily go to the poor rather than to the rich.

Misunderstanding of the Problem of Joint Costs

Often there is also a problem of joint costs (i.e. if a branch line is closed the average cost of the main line station may have been allocated on a proportional basis on the number of trains). However, if the branch line does not exist most of the costs of the main line station would remain but would be allocated over a smaller number of trains. It is also not clear that the concept of the rational consumer is necessarily accurate. If people have been used to travelling by train, but now find that it is less convenient they may make the whole journey by car so that the railways make an increased loss.

Government Intervention and Prices

Part of the problem with the nationalised industries was that in many cases their pricing policy was geared to the economic fashion of the time, in some cases prices were held down, since it was felt that this would have an impact upon the retail price index. On the other hand, if the government felt that it had to balance the budget it would tend to put up fares so that the nationalised industries either made greater profits or at least smaller losses so that the public sector borrowing requirement was smaller.

Similarly, in other industries electricity would be provided in most locations, even if the cost of providing more rural areas might well be very expensive. Often, therefore there was a cross subsidy (i.e. between the rural areas and the urban areas). It is not always clear since the inner urban areas were often poor that there was any economic logic in this. There has never been a suggestion that rural roads should not be provided, even though on most criteria they might not be worthwhile in the really rural areas.

One of the other problems with the nationalised industries was that in many cases they were bound by the ultra vires clause. This means that they were not allowed to expand their activities outside the present ones. This would have been in contrast for example to the private railway companies prior to nationalisation which in many cases had developed for example hotels, road freight companies, buses and even in one case an airline as well as shipping companies so that passengers and freight could complete their entire journeys.

Salaries in the Nationalised Industries

Generally, the salaries at the top of the nationalised industries were below those of the pay in the private sector. However, when Dr Beeching was appointed at a salary of £24,000 per year in the 1960s (when a typical railway salary was under £1,000) having come from Imperial Chemical Industries (ICI), there was a slight uproar. Under privatisation on the other hand the salaries often reflect performance-related pay as well as having free shares or share options. There have been complaints that bonuses have been given, even where targets have not been met or safety controls have been inadequate. Whilst it might be thought that there would be a contrast between public monopoly and private competition in many cases there was not the simple contrast.

Has Monopoly Continued Under Privatisation?

Partly because of the continuing monopoly in many cases there have been controls over prices of the privatised firms. In some cases, the concept of competition has been rather odd for example in the case of the Post Office there have been challenges to the Post Office monopoly, particularly with parcels. However, if the Post Office were to act commercially then clearly it would be much cheaper for it to accept bulk parcels and letters within a town, rather than between for example the South West of England and North East of Scotland. The rolling stock operating companies (ROSCOs), which are responsible for about 90% of the rolling stock, have made monopoly profits, part of the reason they can do this, is that the franchises have been for short periods. No operating company has wanted to buy long-term assets such as locomotives or rolling stock, since what would they do with them if they failed to have their franchise renewed? In 2007, the then Competition Commission announced that it would be carrying out an investigation into the ROSCOs (the train leasing companies).

Assessing the Effects of Privatisation

It is not very easy to work out the effects of privatisation, for example in the case of the railways there have been gains in passenger traffic. How much of this is due to economic growth is a matter of debate. The volume of traffic will tend to rise if the Gross National Product rises. Similarly, if the degree of congestion gets worse, then many people will transfer from cars to trains particularly in the London area where on the whole there are frequent services to London so that overall time will be comparatively less by rail.

This will be even truer when the Queen Elizabeth line, which crosses London between Reading in Berkshire and Shenfield in Essex is completed.

The Queen Elizabeth line had been known as Crossrail. Many transport economists and others hope that Crossrail 2 and Crossrail 3 will also be built.

Types of Price Controls, etc.

One type of control, which could be advocated, would be that of looking at the total profits and putting a cap on these. However, many economists would probably think that this was unreasonable since the organisation might simply to avoid taxation on this give more money to its employees and particular its managers to get around this. If the logic is to prevent monopoly abuse, it is not always obvious why some fares are regulated and others not since in many case the degree of monopoly power is very similar.

Imposing Price Limits and Problems

Imposing price limits has been widely used. In some industries, it was on the basis of inflation minus a percentage each year so that the company had an incentive to hold down its costs. It is sometimes difficult to compare like with like. The fares given in Social Trends are usually those of the maximum or at least standard price which would be charged when in practice many people find cheaper fares for the same journey so that in contrast to many other countries where there is not so great a variety of fares the media often by concentrating on these give the impression that fares are much higher in the country than in other countries whereas this is not always the case.

In the case of the railways it is perhaps more difficult since there are potentially so many different fares between different stations, and in some cases more than one route (e.g. from Paddington to Exeter or Waterloo to Exeter). Since the journey from Paddington is much quicker, it will seem unreasonable to have the same fare for both. They are in any case operated by two different companies.

Measures of Performance

There is also the need to have some sort of measures of performance (e.g. whether trains arrive late, whether they are clean, over-crowded, etc.). However, whilst these may be desirable qualities it is not always clear what is in the public interest. If a train tries to ensure that it is within 5 minutes at the final destination, but this means that passengers at a junction do not catch their train, does this actually help the passengers particularly, if the trains are running an infrequent service? The other possibility which is where economists could adopt a theory of games approach is to see what organisations might do if we are faced with the same provisos and one of these is that if the fines are high they might run slower timetable services so that they gave less compensation to passengers. Therefore, the slowing down of the services would not help the passengers very much. Currently, there have been general criticisms of targets and it is necessary to see whether the targets have an overall favourable effect rather than the opposite.

Who Are the Regulators?

Different bodies (regulators) have been able to set rail fares. The regulators have sometimes allowed for increases in real prices (i.e. above the rate of inflation so that improvements to quality of the railway system can take place). This is a cross subsidy from the present users to the future users. If both present and future users are mainly the same this may not matter. If they are not, then economists would need to consider why this should be done.

Hardship to Users

The 1962 Transport Act specified that there had to be consideration paid to the hardship to the direct users of any passenger lines if a line was to be proposed for closure. Often the hardship was indirect (e.g. the hoteliers and proprietors of guest houses in a seaside town often suffered serious losses if lines to the seaside were shut).

Many of the travellers were probably relatively indifferent as to which of the seaside towns they travelled to. There were suggestions that British Rail got around these provisos by running a very sparse service so that it did not have to be proposed for closure and then when patronage was very low that could they propose the line for closure with relatively few objections.

Freightliner Limited

Freightliner Limited is one of the major container operators in the world and serves a wide variety of ports and terminals within the UK. Much of it is capital intensive though it has experimented with mini terminals such as the one in Bristol. It was privatised in 1996 following a management buyout. It is best known for container load traffic but in 1999 had a new heavy haul division which runs services for aggregates, cement, coal, petrol and runs over 100 services per week. However, Freightliner is now owned by Genesee and Wyoming Inc.

Railways and Land Ownership

British Rail also owned a considerable degree of property. Some of this has been sold where it is no longer used, many private former sidings are no longer required for operational use. Sometimes British Rail developed the property to include major shopping areas as for example with New Street, Birmingham. In 2016, the station was redeveloped yet again and included Tramway links. In 1988, British Rail put forward a massive development project to include both Kings Cross and St. Pancras Stations in London and this has finally come to fruition as St. Pancras became the new terminal for the Channel Tunnel in 2007.

British Rail and Other Companies – Including Sea-Link Hovercraft Ltd and British Transport Hotels Limited

There was controversy at the time about the amount which Sea-Link paid British Rail for the services since the amount was less than the value of the assets. On the other hand, losses were being incurred which no longer had to be paid for by the treasury. Sea-Link under private ownership faced considerable competition not only from rivals, such as P&O, but because it provided a complementary service also from the potential of the Channel Tunnel.

British Rail Hovercraft Limited was owned because it provided a complementary service and was also privatised.

British Transport Hotels Limited was another subsidiary, originally hotels were owned at main termini for example Great Eastern in Liverpool Street, London and Great Northern so that if passengers arrived too late in London to be able to go on to other destinations the hotels were conveniently located. British Transport Hotels Limited also owned much of the catering services including Travellers Fare. In 1988, Travellers Fare was privatised and the management of Travellers Fare remained as one of the potential bidders for management.

Freightliners Limited, another subsidiary was not privatised. Freightliners Limited arose from one of the more positive aspects of the Beeching Report 1963. Up until 1978 National Freight Company had a majority shareholding whilst British Rail had only a minority shareholding in Freightliners. Under the then current system Inter-City was the main user of track, it paid for the cost of the track whilst the provincial services were regarded as the main user of the track so all costs were allocated to the Intercity sector.

Problems of Costing in the Freight Sector

One of the problems for the Freight Sector is that it is not obvious that the heaviest road hauliers still pay sufficiently to cover their social costs in spite of the changes in the vehicle excise duty structure.

There are also passenger transport area local services which received support under Section 20 of the Local Government Act.

British Rail also had a number of subsidiary companies which have now been privatised. These have included Sea-Link UK Ltd., which prior to privatisation operated services to the continent to the Republic of Ireland, Channel Islands and the Isle of Wight as well as minor services such as Tilbury to Gravesend. Additionally, it also owned some harbours and on cross channel routes worked in conjunction with the continental companies such as SNCF.

Part of the reason why British Rail owned the shipping services was that the nineteenth century private operators wanted complementary services which could add to their profits. Often customers preferred to deal with one body rather than several.

British Rail Reorganisation

British Rail often underwent considerable reorganisation. Part of the reason behind the various changes is that it is difficult to allocate costs on any sensible basis because of the high proportion of joint costs and also because there are potential conflicts between operating and commercial decision making. For commercial purposes a regional organisation will often be desirable as customers often want to deal with one group of managers.

For operating purposes, however, there may be problems since a passenger train from Edinburgh to the West of England may pass through a variety of regions and there may be a tendency to pay little attention to operating requirements outside ones' immediate region.

The main structure of British Rail before privatisation was demand orientated. Freight traffic was sub-divided into freight and parcels sectors together with Inter-City services. This formed what might be called a commercial part of British Rail. The Provincial services were those services outside London, which were not considered as part of the Inter-City network together with network South East, which comprises London and South East England.

The Conservative government had reduced the public service obligation grant in real terms and insisted that Inter-City services should break even. Part of the way in which Inter-City services were going to break even is by changes in the allocation of costs. Privatising helped since National Freight was able to diversify logically without the ultra vires clause preventing it diversifying. This enabled it to buy up SPD, the former transport branch of Unilevers which they would not have been able to do if it had remained in private ownership. There was a strong incentive for NFC to be successful, since it was already operating in a highly competitive market. The growth of unit load traffic with higher capital costs particularly for containerisation may have tipped the balance slightly in favour of larger companies.

In 1988, the NFC reverted to a more conventional PLC status partly in order to obtain greater amounts of finance.

The Railways Act 2005

This did several things. It established a railways passenger council which would be a single national body reporting directly to the Secretary of State. This replaced the former regional committees. Not everyone was very pleased about this and it was felt clearly that regional committees may have a greater understanding of the local consequences.

It abolished the Strategic Rail Authority and established a new body called The Office of Rail Regulation, which combined both safety and economic features. The responsibility for railway safety was transferred from the Health and Safety Commission.

The 1984 White Paper on the Bus Market

The 1984 White Paper stated that the bus market was very contestable since there are few barriers to entry or to exit, because of this it was assumed that even if a local bus service had a monopoly, it would not be able to raise prices and reduce supply in the usual way, since otherwise a new entrant would soon enter the industry.

The proponents of the privatisation of the bus companies and the deregulation of the bus companies with the 1980 and the 1985 Transport Acts assumed that there would be many new companies entering the market. Bus patronage with the major exception of the London area has continued to decline and the bus market is dominated by the large bus operators such as Arriva and FirstGroup. Often the present bus operators also have a rail interest.

Scottish Bus Group

The Scottish Bus Group was established in 1968. Whilst the bus interests were predominant it also had both shipping and road freight interests. The Scottish Bus Group had, as with the National Bus Company, split itself into a variety of regional subsidiaries covering the various parts of Scotland mainly most of the densely-populated parts of Glasgow, Dundee, Edinburgh and Aberdeen. Its shipping company is Caledonian MacBrayne Ltd. The road haulage company has been MacBrayne Haulage Ltd. which has operated mainly in the Western Highlands and Islands. There was controversy in 1988 about whether the Scottish Bus Company should be sold as a single operator or split up into smaller parts. Eventually, it was decided that it should not be sold as a single entity but split up.

British Airports Authority

British Airports Authority (now Heathrow Airport Holdings) replaced the former control by the civil service, it was eventually privatised and was bought by a Spanish firm Ferrovial in 2006. In 2009, the company was required to sell Gatwick and Stansted airports, and over the following years sold all its airports other than Heathrow. The company was renamed Heathrow Airport Holdings in 2012 to reflect its main business.

Access to Airports

An extension of the Piccadilly line connected Heathrow in 1977. This has been successful though the majority of people using the connection either join it at a Piccadilly line station or at a District line station which has a cross platform connection with the Piccadilly line. The Department for Transport announced in 1988 that it wished to see a mainline connection from Paddington and this has now been built as the Heathrow Express. In 2009, the government announced that a third runway would go ahead as well as a new railway line from Heathrow to St. Pancras and Birmingham.

While Gatwick Airport is further out from London, it has a better rail connection from London Bridge and to Victoria, and with the subsequent Thames Link line being reopened it has links to Kings Cross as well as to the Bedford line. There has been a dedicated Gatwick Express which has general charged higher fares than the alternative. There was considerable discussion about this in 2007. The building of the M25 has improved the road links.

Stansted has had only a relatively small number of passengers until recently. It is 35 miles from London and it now has a direct rail service to Liverpool Street which is quick and there are also direct routes from many other parts of the country including the Midlands.

Airport Ownership

The former British Airports Authority (BAA) also owned Aberdeen Airport. Demand for this grew substantially with the oil boom in Scotland. It has been widely used both for flights to London but also for oil charter to Shetland.

In the mid-1980s, BAA was privatised. It faced a very limited amount of competition within the London area from City Airport, located within the London Docklands. It might be expected to diversify these activities into other terminal activities (e.g. expansion of hotels near the major airports). The single market 1992 abolished duty free for EU visitors. BAA gained considerable revenue from this.

Originally, the airports subsequently controlled by BAAs were directly controlled by Civil Servants in the former Ministry of Transport and Civil Aviation. The department had a variety of different names in the post war period. After the 1961 White Paper, state owned airports were transferred to the British Airports Authority. One of the aims was that the airports should become commercially self-sufficient.

One of the main problems which BAA had to contend with is whether to build a third London airport which was originally going to be at Stansted then at Maplin and subsequently Stansted has become the third London airport. It also has the problem of trying to expand capacity at Heathrow. The fourth terminal at Heathrow was eventually opened in 1985 at a cost of about £200 m and increased overall capacity there. The airlines wanted a fifth terminal which has now been opened and which has access both from the M25 and also another extension of the Piccadilly line. The advantages for the airlines would be that they would not have to duplicate flights and that passengers would not lose in travelling time. BAA on the other hand was reluctant to expand, partly one suspects because of likely environmental objections.

British Airways

After the Second World War two major public corporations were set up. These were British European Airways which served Europe and British Overseas Airways (BOAC) which operated on the longer routes. Whilst most of the routes were run for commercial reasons there were some social service routes including those to the Scottish Highlands and Islands and also to the Isles of Scilly off Cornwall.

In 1968, the Edward's Committee which had been set up by the then Labour government recommended that there should be a merger between the two airlines. In 1973, the two airlines were merged into a new organisation called British Airways. The main reason for the merger was that it would gain from economies of scale aircraft flying from Britain to North Africa would be similar to those flying to Eastern Europe and so there would be economies of scale in maintenance, spare parts as well as advertising and marketing and the route structure could be rationalised. However, there were also some potential disadvantages since the two airlines were geared generally to different needs and therefore marketing had to be somewhat different and that there were also the usual problems of potential jealousy.

The Edwards Committee had also suggested that there was need for an independent airline. It was suggested that this could be helpful particularly in trying to boost Britain share of the market on the UK to North America routes.

British Airways was privatised in the mid-1980s partly because it was felt by the then Conservative government that this would lead to increased competition. However, in 1987 there was considerable controversy over its take-over of British Caledonian since it was felt that this would lead to a greater degree of monopoly on domestic routes. On international routes, there is a considerable degree of international competition.

In May 2017, British Airways suffered from a computer glitch which affected many airlines as well as British Airways. Whilst this was a technical fault, the lack of staffing at terminals at both Gatwick and Heathrow led to widespread bad publicity. It was estimated that direct compensation would cost around £150,000,000, but the overall costs to British Airways could be much higher since their poor treatment of passengers would lead to loss of revenue in the future.

The British Waterways Board and the Inland Waterways

The British Waterways Board which runs a great percentage of the inland waterways but was set up under the 1968 Transport Act. Under the Transport Act 1968 there are three types of Inland Waterways. There are those which are used for commercial purposes only for the carriage of freight, those which are used mainly for cruising and for recreational purposes and the remainder.

The commercial waterways amount to just over 350 miles and cruising to just over 1,000 miles. There been an increase in the use of commercial waterways though these have not generally covered their costs. The Select Committee on waterways in 1978 suggested that there should be greater use of Inland Waterways. One of the major developments in recent years has been the Sheffield and Yorkshire scheme which received financial assistance from the then EEC as well as from South Yorkshire County Council. It permitted much larger payloads than had been previously carried. The maximum potential payload is about 700 tonnes. There has been much less interest in carriage of goods by Inland Waterways in this country than in other countries though this is partly because of geography, the Rhine is a much deeper and longer river.

The National Freight Corporation

The National Freight Corporation was set up under the Transport Act 1968. The 1947 Transport Act had nationalised most of the hire and reward sector though there had been partial denationalisation in 1953. After the abolition of the British Transport Commission in 1962 the Transport Holding Company had both Road Freight and Road Passenger interests.

The National Freight Corporation as it was called before privatisation controlled a great number of well-known road haulage operators. This included the British Railways sundries division which had made larger scale losses whilst in British Rail ownership. It also included British Road Services as well as Road Line UK Group which had previously been known as British Road Services Parcels Limited. It also included Pickfords, which was well-known for removals work.

Unlike many of the other nationalised industries National Freight Corporation only accounted for a relatively small percentage of the total market about 10%.

The denationalisation of the National Freight Corporation which subsequently became the National Freight Consortium was successful. This may have been partially because it was a management buyout with both employees and pensioners encouraged to buy shares in the company. This helped to motivate workers who no longer felt alienated so much from management and were no longer subject to the whims of government policy.

National Bus Company

Originally, a great number of bus companies were nationalised in the 1940s partly because many bus companies had been taken over by the railways and partly because a number of subsidiary companies of the Tilling Group were taken over in 1948. Under the 1962 Transport Act a Transport Holding Company was formed which took over a variety of companies including what subsequently became the National Bus Company as well as the National Freight Company. Under the 1968 Transport Act the National Bus Company was set up. Whilst the National Bus Company had overall financial control of bus companies the different bus companies continued for operating purposes. In the Midlands, the Midland Red Company was well-known as in the South East region were Eastern National, Eastern Counties, etc.

The National Bus Company was also responsible for the National Express network which was more an organisational name than an operating unit since it mainly used the coaches from the individual companies.

Finance for the National Bus Company came partly through fares, but also from grants from Central Government towards new buses until 1983.

To qualify for this grant, buses had to be used for at least half their mileage as a stage carriage service.

Under the 1972 Local Government Act which re-organised local authorities, counties could provide the cost of bus services for social reasons and substantial sums were paid for this.

CHAPTER 17
Cost Benefit Analysis

Why Is There a Need for Cost Benefit Analysis?

Cost Benefit Analysis (CBA) is widely used in the evaluation of transport policy by the Department for Transport. If the pricing system adequately measured social costs and benefits, there would be less need for a CBA. The EU has stated that roads and other modes of transport should use a marginal social cost pricing system. The congestion charge in London partially does this and if this could be further refined then there would be much less need for CBA.

Evaluation of Externalities

However, currently there are many externalities which can be defined as costs or benefits of transport services which are not directly paid for by the consumers or producers of the transport services. These have conventionally included increased or decreased employment, noise, pollution, accidents, severance, congestion and vibration. In the 1980s when unemployment rose to 3 million in the UK, there was a considerable difference between the opportunity costs of labour to the firm and to the community. Even with the lower rates of unemployment in the UK currently there are still considerable differences and this applies even more in many parts of the EU where unemployment is still high.

In the UK, economists could look further than this, then the firm will be paying not merely the wages but also the national insurance employers' contribution. Therefore, if the person was earning £20,000 the total cost to the firm ignoring the cost of the office, etc. would be £21,400. On the other hand, if the person was not employed, the government would not be gaining either the employees or employers contribution, would not be gaining the value-added tax from the purchases which the person would make and would also be paying Social Security. We could tabulate this as follows (Table 17.1).

Table 17.1 Two types of financial accounts

Expense Item	Financial Cost to Government	Net Cost or Benefit to Government
Salary of Worker	£20,000	£20,000 cost
NI Employer Contribution	£1,400	Zero
NI Employee Contribution	Zero	£1,400 benefit
Additional Overheads (e.g. office)	£1,000	£1,000

In addition, some external costs may be very high. In the mid-1980s, it was estimated that road accidents cost about £2 billion per year. Since then the number of road deaths has been reduced considerably. The costs would include the time off taken by people as a result of accidents. It would also include the costs to the National Health Service as well as the loss of production. The method of valuation is discussed later on.

The Channel Tunnel and Employment

Partly because of the high costs of unemployment, many commentators from the trade unions and the Confederation of British Industry at the time suggested that there was some scope for increasing public sector investments in the transport infrastructure. The Channel Tunnel group in their literature about the then proposed Channel Tunnel in 1985 emphasised the employment that would take place as a result of a major construction. Economists would wish to look at both the short and long run effects. During the construction stage, many people would have been employed before the tunnel opened in 1994. Once the channel tunnel had opened economists would have assumed that there would be been some job losses in the ports of Dover and Folkestone.

Example of CBA

One of the best known early examples of CBA was the M1 motorway appraisal in the early 1960s. This was a rather odd use of CBA since it was conducted after the motorway had already been built. Because road construction is not paid for directly, the main benefit was journey time. There are also other benefits such as a reduction in accidents.

The Victoria Line, the Jubilee Line Extension and the Elizabeth Line

The original assumption when the Victoria line was first considered was that it would be profitable. It was built in 1968 linking among other stations Victoria and Kings Cross and Euston. When it was realised that it was unlikely to be profitable the CBA took into account time savings to users who would not need to make several changes from Victoria to Kings Cross. There were also benefits to other users on the Underground lines since these lines were less congested. The use of consumer surplus is extremely important and it reflects the difference between the price that customers pay and the price which they would be prepared to pay because of a better service. Since the fares were no more expensive on the Victoria line than for the same journey before it was built the consumer surplus was considerable. The Victoria line cost benefit analysis also showed that road users gained from the construction as there was less road congestion once the line was built. More recently the extension of the Jubilee line to Stratford has meant that far more people now travel by public transport than by car to the Docklands and there is therefore less congestion because of this. The cost benefit analysis on the Victoria line distinguished between the different types of time which were being saved. Generally, business people would put a far higher valuation on their time than people travelling for social purposes. Transport economists can observe who is willing to pay for taxis which are faster than buses. The Elizabeth line opening in 2018 has had many people trying to evaluate the line on a cost benefit analysis approach.

Severance

Severance is where one section of the community is severed from another because of the building of a new transport service (e.g. a new road may sever one part of a farm from another). Sometimes people will not be able to cross from one side of a road to another so that new schools may have to be built as a result of new road which means it is no longer safe to use it. Sometimes as with the then new Severn bridge in the 1960s a transport project may have the opposite effect so that communities which previously had to have two sets of amenities may be able to enjoy one set.

Roskill Commission on the Third London Airport

The Roskill Commission on the third London airport in the 1970s was an example of cost benefit analysis. The terms of reference for the Commission has been criticised since many people felt it was not asked to consider the right questions. It was not allowed to look at the expansion of the existing airports such as Heathrow and Gatwick or the capacities of what were then called the minor airports such as Stansted, Southend and Luton. It was also not allowed to look at the "do nothing" option and just had to decide between the proposed airport sites. The commission at one stage tried to make enquiries about the cost of expanding existing airports, rather than building new ones but was not allowed to do this. This is ironic since in the mid-1980s the Eyre report recommended the expansion of Stansted and this has taken place. Also since that time, terminal 5 has now been built at Heathrow Airport. This however, is more a criticism of an investment appraisal rather than a criticism of cost benefit analysis. It does highlight the point that cost benefit analysis is a planning tool rather than an entirely objective method of assessment. This is partly because there is no universally accepted method of quantifying social costs and benefits. Some costs may be relatively easy to determine by looking at the price mechanism. Economists can look at prices of properties with and without transport development to see what gains take place.

Measuring Externalities

In principle, there are two main methods of trying to measure external costs and benefits. The first is the willingness to pay principle and the second is the compensation principle. Economists could ask people how much they would be willing to pay to receive a particular benefit whether it was a quicker journey by car, coach or train or work or holiday or the avoidance of more noise or spoiling of their view from their houses. It would be possible to use questionnaires but it would be difficult to obtain honest reliable answers since most people are not used to putting numbers into what they would regard as a hypothetical case. Other people assume that absence of noise is one of the rights which they should have. Better valuations could be obtained by looking at houses which are otherwise identical apart from noise or possibly pollution. This would be easier to do with car noise rather than air noise which tends to cover a much wider area. With the compensation method, it is more difficult to get an adequate answer.

The two methods may well give different answers, for example, if we ask people how much they will pay to avoid risks of fatal accidents then clearly there is an upper limit which is imposed by the income of the people being questioned. If we try to measure through the compensation principle, then clearly most people would require a very high figure as a compensation for their families. Adopting the second method might well give a more realistic assessment than the very low figures which have been used in the past by many economists and politicians. The first method assumes that the existing income distribution is satisfactory.

If a new road scheme were to be suggested, rich people might be willing to pay much more to eliminate the effects of heavy traffic than poorer people. This would not necessarily imply that rich people suffer more from noise but merely that they have more income or wealth. Some economists have suggested that it is difficult to make inter-personal comparisons, though the price mechanism, as usually applied, does this. The importance of making inter-personal comparisons is that many measures would help some people, but would harm others. A new airport would benefit customers but would be a nuisance to nearby residents. A new road scheme, such as the proposed one near the Archway North London, would benefit road haulers and motorists using a very congested section of road, but would not be welcomed by many local residents whose houses might need to be destroyed. Partly because of this, this particular problem has been a continuous source of anxiety in the UK for many years.

The second approach, using the compensation principle, would value poor and rich people's ideas at much the same values.

Criticism of CBA

A more valid criticism of CBA is that it often ignores who benefits and who pays. A motorway through a poor area means increased noise and pollution to local residents but possibly great savings in time to the more affluent motorist using it. By the first method willingness to pay would show that the benefits outweigh the cost, whilst the compensation principle might lead us to the opposite conclusion. Since much transport infra-structure benefits the rich rather than the poor, the unintentional effect of cost benefit analysis would be regressive (i.e. the burden of bearing the social costs would fall more heavily on the poor).

Alternative Methods of Funding for Externalities

Some people would wish to see a site valuation tax imposed upon the people who benefit from a new transport development. Some benefits such as peace and quiet or the benefits derived from open spaces or historic buildings are more difficult to put into monetary terms.

Valuing Time Savings

Benefits often include time savings, particularly with many road building schemes. This saving is usually sub-divided by economists into working and non-working time. Observation suggests that most people value savings in working time more than non-working time. Thus, people might use a taxi for business purposes to save working time, but would not necessarily consider using a taxi to save leisure time. People will be more willing to take a taxi if their firm is paying than if they have to pay themselves, it is still observed that people are more concerned about saving working time than leisure time from a purely time saving aspect, even though there does not seem to be much logic in this behaviour.

Working time does not include time spent commuting. The value of short periods of time is not necessarily the same as with longer periods. This is partly because we cannot always use say an extra two or three minutes which might be the typical time saving with improving a junction whereas we can use a period of say 2 hours which might result as a result of a new international transport route.

The Balance Sheet Approach

It has sometimes been suggested that rather than trying to put what are said to be spurious price values on total different idea such as accidents, time savings, peace and quiet that instead we should simply have a balance sheet approach which shows the various benefits and disbenefits so that we can judge more easily whether we want to go ahead with a project or not.

CHAPTER 18
Local Transport

Local Government and Road Maintenance

There has been a long history of Local government involvement in transport within the UK, for example, local authorities were responsible for the roads in the sixteenth century because of the 1555 Act. Under the 1870 Tramways Act local authorities could set up tramways if they had the consent of the highway authority, if two thirds or more of the highway was used by the trams other highway authorities had to agree.

Local Authorities and Trams

Local authorities tended to be involved indirectly, because they had compulsory purchase options after 21 years and then at 7 year intervals. The tramway operators therefore had little incentive to electrify the tram system and therefore in many cases the local authorities took over the tramway operations. Later as the trams with the exception of Blackpool were taken over by trolley buses and diesel buses, the local authorities often became bus operators. The trams in many cases had the disadvantages of having to pay for the cost of their tracks and the road within 18 inches on either side without the advantages of the railways, which paid for their tracks but had exclusive running except on level crossings and even here the railway had priority. There were some minor exceptions to this rule (e.g. in Weymouth the railways to the port did run along the roads). The tram did not have these advantages. There are now however tram systems in several towns including the one from Croydon to Wimbledon etc. in the London area as well as in Sheffield, Manchester, Nottingham and the West Midlands with the Midland Metro. Edinburgh has plans for a tram system using much of the old suburban rail network trackbed.

1974 Changes to Local Authorities

Until the changes in 1974, only boroughs had been transport authorities and county councils had not been involved directly as operators. After 1974, the Passenger Transport Executives had the same area as the county councils and therefore in most of the conurbations the county councils had indirect involvement in transport provision. Outside the major conurbations, many local authorities such as Southend-on-Sea continued as bus operators.

Watershed Areas

One of the problems with local authorities running transport services had frequently been their jealousy over allowing other bus services to run within their area. This frequently gave problems in the so-called watershed areas (i.e. those on the fringe of two municipal authorities). The problems of watershed areas were not however confined solely to operations run by local authorities.

The London Area

Prior to the changes in 1974, there had been a somewhat similar change in London where the London Government Act 1963 created the Greater London Council (GLC). As the overall planning authority. Thus, the GLC became a highway authority for 560 miles of metropolitan roads and was a traffic authority for other roads apart from trunk roads. The Transport (London) Act 1969 transferred control of London Transport from it being a nationalised industry to being a local authority one. This subsequently changed and London Transport reverted to being a nationalised industry.

In 1986, the local authority system changed yet again in the conurbations and outside the conurbations due to separate legislation, the district councils had an arm's length relationship with their bus undertakings.

In the London area, Transport for London (TfL) is controlled by the elected Mayor of London and Ken Livingstone whilst Mayor of London, introduced the congestion area, Boris Johnson, whilst Mayor of London bought in more cycle lanes, as well as so-called Boris bikes/Santander Cycles (formerly Barclays Cycle Hire) which could be hired at over 750 docking stations across London.

The current Mayor of London Sadiq Khan is concerned about pollution, particularly in the Oxford Street area.

Transport in the Conurbations

In several conurbations, there are now elected mayors who have responsibility for both transport and planning. Some of these are high profile for example, Andy Burnham, the former Labour Cabinet Minister.

Land Use Planning in Other Countries

The nature of transport means that it uses large amounts of land for the infrastructure. This would include therefore both the highway system but also the parking lots and car parks. Even in a slightly less car orientated country such as the UK in the suburban areas the houses themselves often take up less land space than the roads which serve them. Also, the ways in which land is used determines partially the type of transport system. For example, if the population and activities are uniformly spread it would be more difficult for fixed track systems, which consists of trolley bus, tram and rail to provide a service than where the activities are heavily concentrated.

Originally much of the transport demand and infrastructure was influenced by the location of natural resources. The West African rail network was often used by the colonial powers of the day to obtain raw materials and bring them to the coastline. The narrow gauge was used in countries, such as Nigeria, since the demand for such railways services in difficult terrain was often cheaper if such services were built to this standard rather than to standard gauge which would have been costlier. This has had repercussions (e.g. a recent Nigerian railway report referred to this). It made it more expensive to buy rolling stock and locomotives, etc. from many firms since there are obvious economies of scale, in which manufacturers of equipment can gain with items of the same gauge.

Major events such as the Olympics and Paralympics in Brazil in 2016 meant that the land use planning had to be altered very substantially in order to cater for the increased number of visitors.

Narrow gauge was often used for mineral traffic in the UK. In the UK, a great deal of the original railway system even before the introduction of steam locomotion was for the carriage of coal. If coal could not be transported cheaply, then it was often sensible to locate heavy industries near the coalfields. With the increasing use of electricity which is virtually countrywide this has meant that industries have become much more footloose. Oil is relatively cheaper to transport than coal and has also made industry more footloose.

The main current reasons for the interest in land use and transport is probably because of externalities such as accidents, pollution, noise, severance of communities, visual intrusion, congestion as well as the large amount of land each transport mode uses.

Predict and Provide Comparisons of the US and the UK

In the UK demand for roads, etc. was often on a predict and provide basis. This followed from land use transport surveys (LUTS) often used in the USA. However, this method can be criticised on many grounds. The first is that whereas the US has high car ownership the amount of land in the USA is far larger per head of population so that even if the system is good for the US, it is less likely to be good in the UK or other densely populated countries.

The second is that whereas much of the urban parts of the US are laid out on a grid system which may be suitable for newer areas, the UK has many towns where the character of the place would be completely destroyed by such a system. It is doubtful whether for example the residents of towns such as Norwich or Oxford would appreciate having parts of the town centre demolished to make way for more cars.

In the case of airports, once the airport is there on the basis of predicted demand it is very difficult to move the site so the irreversibility of decisions has to be borne in mind.

The effects of town planning both on total demand and on modal split has become more recognised. At the present time, town planners are supposed to look not just at the typical planning policies (i.e. whether one building or use will affect another building/use/occupier use as one of the main aspects but also the consequences for the environment, including the transport consequences).

Land Use Planning

Planning can be regarded as a collective want. This phrase means that one person cannot generally determine the nature of his or her demand in isolation from other peoples' demands. In a free market, each person might try to live as near to the countryside or to beauty spots as possible. A newcomer will therefore try to build a house which takes up part of the existing countryside. It is also difficult to find out through either existing price or administrative mechanisms to know whether people would prefer different densities of housing with a different range of transport options. This has been very noticeable with council housing particularly the range of tower blocks which were popular amongst planners at least in the 1960s, but are now being regarded as undesirable.

Local Authorities and Land Use Planning

Government guidance on transport planning is contained in Planning Policy Guidance Note 13 Transport. However, one of the issues in the UK is that there is no National Transport Strategy with the exception of the Wales Transport Strategy. Transport issues can clearly impact on more than one region and therefore the lack of a National Transport Strategy has made it more difficult for decision makers to consider the impact of their decision on wider geographical areas.

This can also be seen at a more local geographical area. Prior to the Local Authority changes in 1974 there were often problems of combining transport and land use planning in conurbations. In the West Midlands, there are a large number of different local authorities such as Walsall, Wolverhampton, Birmingham, Coventry, etc. This meant that it was difficult to have any overall planning to ensure that factories in one local authority did not have unfavourable externalities for residential housing in another authority. Prior to 1968 there was no overall transport authority in most of the major conurbations though the advent of the Passenger Transport Executives overcame part of this problem. One of the criticisms which led to the development of the new local authorities was that there had been little integration of land use plans with transport.

Planning and Compulsory Purchase Act 2004

The Planning and Compulsory Purchase Act 2004 introduced various changes to the planning system. Local Planning Authorities will prepare local development documents (LDDs). These will replace Structure Plans, Unitary Development Plans and Local Plans.

Forecasting of Demand for Transport

The data which will be required for all such plans would be data about employment which is very difficult to forecast in a period of rapid changes, caused partially by the digital revolution and also with rapid changes caused by high levels of unemployment for example following the credit crunch 2007 onwards. Any plans will need to take account of past and proposed changes to housing and other buildings such as new employment uses and consider what impact there would be on transport.

One of the problems in formulating Structure Plans was the problem of data. The proposals for a mini census were abolished which meant that less data is available to local authorities.

The 2011 census has provided a wealth of information to transport operators and local authorities as well as the Department for Transport (DfT).

Over Optimism About Economic Growth

Local government nearly always overestimated potential economic growth and therefore any local authorities who had based their forecast on such a system would have tended to over provide transport facilities.

Highway Engineering Bias

One of the criticisms which was made of the local authority system was that most of the planners had a highway engineering bias and therefore insufficient attention was paid to using other modes of transport. Perhaps because most of the decisions are made by middle aged car owners too little attention has been given to walking or cycling though both modes can be important.

Even though the metropolitan counties were abolished in 1986, there will still be a need particularly in conurbations to have models of passenger demand.

Current Changes

Currently in the UK the new metropolitan mayors in six conurbations will be interested in new developments across their areas.

Modelling Passenger Demand

When modelling passenger demand, the main variables such as incomes, household size, age of the population, car ownership will need to be taken into account. All of these influence demand, but are difficult to forecast. Data can be used on either a time series or a cross section data basis. Economists can obtain data over time showing traffic trends and use regression methods. One of the problems with this is that it assumes that past performance is a guide to the future. This may not necessarily be the case if there are shocks to the system as with Oil Petroleum Export Countries (OPEC) price rises in 1973–1974 or the halving of oil prices in 2015 the credit crunch 2008 or British withdrawal from the EU following the referendum in 2016. It may also not be a guide if there are potential constraints, roads may become increasingly congested and therefore may deter the future traffic growth. The other way of using data is through cross section data, economists may assume that different income groups or certain economic groups may have different patterns of transport behaviour. It is important when looking at data to recognise the differences between correlation and causality correlation shows an arithmetic relationship between two variables. It might be assumed that there is a positive correlation between incomes and car ownership (i.e. that if one rises so will the other). A negative correlation exists between price and demand, if price falls we would expect demand to rise.

Forecasting of Demand and Robustness

For some purposes, the degree of accuracy may not need to be very great, if we only need to know whether to build a road, there may be a wide range of traffic which would warrant the building of such a highway. The same may apply to the number of lanes. For public transport, however bus companies may need more accurate data since for example bus companies will need to know at what time the traffic is likely to be there and also the particular location.

There are two main ways of using models, economists can either use aggregate data (i.e. they can build up models from individual data or they can take global figures). If they use global figures. They may not notice that there are some underlying trends which are likely to alter the trends in the future. Individual data can be helpful, although we might not always notice the relationship between different strands (i.e. when the Jubilee line extension was built to Stratford in East London and the extension and improvements to London Overground would have altered not only the Underground data but also road transport as well).

One of the problems of forecasting transport demand is that there may be a frustrated demand particularly for public transport which is not being met. Surveys by county councils have sometimes discovered that people would like to visit certain towns for shopping but there is no bus service for this, even though it might be profitable.

Use of the Price Mechanism

The price mechanism can be used to some extent. If we had road pricing perhaps not only for urban congestion but also for newer roads it might give a much clearer indication of how far people wanted roads rather than relying upon the weights given to the various pressure groups pressing for roads on the one hand and the environmental groups often campaigning against roads on the other. It might also be able to be used if for example passenger transport undertakings could buy priorities for their vehicles by in effect purchasing part of the road.

The London Docklands Development Corporation 1981 to 1998

The link between transport and development in this area is an interesting one. The London Docklands Development Corporation (LDDC) as the name implies originally catered for the wide variety of shipping that came up stream to the Thames especially in the nineteenth and the early part of the twentieth century. Since that period there has been a general tendency to move downstream so that the majority of shipping later transferred to Tilbury Docks. In turn, most River Thames shipping now uses either Thames Haven in Essex or Thamesport on the Isle of Grain in Kent. The London Docklands have 321 miles of river frontage and 10 miles of quays within the enclosed docks. The majority of the river views have been hidden from sight by the raised banks and often disused wharves. The main financial attraction for new firms moving into the LDDC was the relaxation on many planning controls and perhaps more importantly the absence of rates for the first 10 years. This was also combined with its accessibility to London which has been rapidly improved by the Docklands Railway and also the short take-off and landing (STOL) airport.

East London River Crossing

There have been also proposals for an East London river crossing which would cross the Thames from the A13 North of the Thames to the A2 South of the Thames. This was originally to take the form of a bridge but because of environmental objections would be likely now to be a tunnel. However, not all neighbouring boroughs would be in favour of this and Greenwich Council has suggested that there could be improved links by the use of a third Rotherhithe Tunnel which would minimise environmental complaints.

The corporation was also interested in improving the passenger links between Central London, Greenwich and the Docklands Corporation area. One of the difficulties has been the speed limits which have been imposed on craft using the Thames because of problems from excessive wash. In 1988, a new riverside service opened with speeds up to 25 knots. However, this was discontinued partly because of problems of debris on the River Thames which disrupted the craft used.

Reclaiming Derelict Land

One of the major problems for the corporation was reclaiming derelict land and providing a more attractive landscape. One of the failures of the existing pricing scheme is that firms may leave land in a derelict state which may be profitable for them, but may harm the community and that other firms or potential house dwellers will be unlikely to enter such an area. Partly for this reason therefore the LDDC was given the powers of compulsory purchase and has in many cases improved the landscape, for example from the former Gas Board spoil in Beckton. This is no longer required as an area because of the development of natural gas. The possibility of a Marina Centre which might include boat exhibitions was also considered.

There is now a marina at St Katherine docks. Old warehouses have often been converted into a variety of other purposes including sports and leisure purposes. These again can if properly converted be quite attractive though in their derelict state will be unlikely to attract anyone.

A large number of firms have moved into the area including the Daily Telegraph, the Guardian newspapers and education facilities have been improved with Greenwich University opening up a large new centre. Currently the Thames Gateway corridor is being developed on both sides of the river.

Lack of Continuity of Control

One of the problems with local authority control of transport undertakings whether buses, ports or airports is that there may be lack of continuity of control.

This is particularly likely where there are annual elections and even where in other local authorities, elections are held every four years, this may be too short a time for long term investment. County councils do not provide transport services, but are highway authorities and can subsidise schemes where they would not be provided on commercial terms.

Political Differences

There have been political differences between the political parties with the Conservatives being perhaps generally less willing to subsidise public transport and being biased more towards highway development. The Liberal Democrat Party and the Labour Party on the other band have tended to favour more development of public transport. Because of the coalition between the Ulster Democrats and the Conservative Party following the 2017 general election, the transport structure in Northern Ireland will be very important.

Political differences have been particularly important where as in 2007 many county councils had no overall control and it has therefore sometimes been difficult to have an overall policy towards transport and planning. Whilst local authorities have sometimes been accused of lacking transport expertise this criticism seems slightly harsh, it is doubtful if shareholders in public or private companies have any greater knowledge and the degree of management expertise in many transport undertakings leaves much to be desired.

There seems to have been more valid criticisms by a variety of commentators. One of these is that of transport being a political shuttlecock with changes of political control making long term planning difficult. This has been particularly noticeable in the conurbations where if bus garages are demolished or railway lines closed it is difficult to reinstate them.

Whilst both the buses and the Underground have been within control of London Transport there was relatively little evidence of co-ordination until the 1970s when a common system of ticketing has been used. TfL also does not control the rail services which are particularly important South of the river where there are currently few Underground services although the East London line has been extended from its rather odd route from Whitechapel to New Cross and New Cross Gate. The Oyster card scheme has been useful in the London area to reduce administration costs and these savings have been passed on to passengers. The Stratford North – Woolwich part of the line from North Woolwich to Richmond has been taken over by an extension of the Docklands Light Railway (DLR). The Docklands Light Railway has had several extensions with that to the Bank in 1991, to Beckton in 1994 to Lewisham in 1999 to London City Airport in 2005 and the extension to Woolwich opened in 2009 and to Stratford International in 2010, and currently there are plans to extend further. At Stratford International station, there are links to the fast domestic services from Kent to St. Pancras.

The Jubilee line was greatly over budget when it was finally extended to Stratford just in time for the Millennium Dome Exhibition.

London Transport History

In the 1920s, road passenger transport in London had been dominated by a few large organisations. The London Passenger Transport Board was set up in 1933 as a nationalised industry. The board members were appointed by the Minister of Transport and this pattern for the nationalised industry was continued with the widespread nationalisation of many industries in the 1940s. One of the reasons why London Passenger Transport Board was not directly under the control of the then London County Council (LCC) was that the area covered a large part of other counties including Middlesex, Surrey, Hertfordshire, Essex and Kent.

The London Transport Executive

In 1948, the name of the organisation was changed to the London Transport Executive. In London Transport from 1933 onwards controlled both buses and Underground and this was continued when the Transport (London) Act 1969 revised the structure and transferred its powers to the GLC. The Greater London Council (GLC) had been set up in 1964 and covered a much wider area than the former LCC. The country buses of London Transport and Greenline Coaches which mainly served outside the GLC area were however transferred to London Country Bus Services and this became a subsidiary of the National Bus Company. Now transport comes under the Mayor of London and there is overall control under the Transport for London aegis. In July 2003, the London Underground network came under the control of Transport for London.

Passenger Transport Executives

These were originally set up under the 1968 Transport Act to provide an overall passenger transport pattern for the conurbations. This was before the majority of conurbations except London and the idea of one overall local authority had been set up. In 1972, there was a Local Government Act and the Passenger Transport Executive Areas were the same as those of the Local Government areas. The Passenger Transport Executives comprised Greater Manchester, Merseyside, Tyne and Wear and the West Midlands. After this two other Passenger Transport Executives came into being comprising South Yorkshire and West Yorkshire and in Scotland there is a Passenger Transport Executive for Greater Glasgow. The Passenger Transport Executives have been both bus operators which have been very large in the Greater Manchester area and sometimes rail operators. They have also run railway services notably the revitalised Glasgow Underground and the Tyne and Wear Metro as well as trying to improve facilities together with former British Rail for the provision of local rail services. They have been abolished but control of some transport and planning services is now under the control of the Metro Mayors.

The Advantages of the Tyneside Metro System

For a long while the standard British Rail services had faced a typical vicious circle (i.e. that demand was declining and so there were likely to be cutbacks which in turn gave a poorer pattern of services and so the decline continued). The disadvantage of reducing the service by closing stations or reducing the number of trains was that it would have lengthened the accessibility time too considerably since most existing passengers walked to the station. If the trains were cut it would have lengthened waiting time.

The disadvantage of using the railway as a bus lane was that because of the narrow width of the track there would have been restrictions at certain points which would have reduced bus lanes considerably. It would also have been difficult to have given accessibility into the town centre. To have closed the railways and used ordinary bus services would have added to congestion in the centre of the conurbation.

Here the committee had to decide whether to run down the existing diesel rail service which was currently making a large loss or to leave it as it was or possibly try to convert to bus roads as in Runcorn. Eventually the committee decided that a metro system mainly using its existing track was the most desirable. The existing North Tyne loop railway was extended across the centre of Newcastle whilst another British Railway suburban railway to South Shields was improved and made part of the metro system. There has also been an extension to Sunderland.

Many of the bus routes were diverted to provide a feeder service to the metro stations. The new services were successful, partly because they were very frequent.

The West Midlands Passenger Transport Executive and Subsequent Changes

The West Midlands Passenger Transport Executive (PTE) on the other hand had decided on a light rail transit system using light weight vehicles. It claimed that as with the Tyne and Wear Metro that it improves the environment and the articulated rail cars carrying 170 passengers that it will provide safe and speedy travel. The first route which the West Midlands PTE opened in 1992 used 12 miles of disused track between Wolverhampton and Birmingham. This follows the former Wolverhampton to Birmingham Snowhill route. They both serve over 20 stations and served parts of the West Midlands which have not had a rail link for a considerable time. This is important attracting through the multiplier further investments into the region. Subsequently, there have been changes within the West Midlands and a tram system has been set up with a very frequent service to Birmingham New Street station.

Concessionary Fares Systems

A great number of councils also gave support towards concessionary fares systems. These were not confined to the National Bus Company (NBC) though in many of the shire counties in which they operated the National Bus Company would have been the major operator. It also varied very considerably Oxfordshire being one of the counties giving the least support whilst Clwyd gave considerably more support.

In some metropolitan counties, the NBC operator had an agency agreement with the PTE. Sometimes, as in the West Midlands joint services were run on some routes with the Passenger Transport Executive.

In the late 1980s, the NBC was privatised. Instead of being sold as one unit which might have created unacceptable private monopoly power the various parts were sold. Sometimes there were management/worker buyouts whilst sometimes the former NBC subsidiaries were bought out by other operators. There are now free bus fares for passengers over a certain age in the off peak. The age was originally 60 but has subsequently been raised slightly.

New Towns Act 1946 and Subsequent Developments

Under the New Towns Act 1946 the Minister could appoint a development corporation which could acquire land and provide buildings and services. There have been two main types of new towns those which have been built around the London area, for example Basildon, Crawley, Harlow and Stevenage which have been developed in order to avoid some of the problems of over congestion in the London area. There have been other new towns such as Washington in Tyne and Wear and Cwmbran in South Wales which have been built in development areas to provide a focal point for those areas. In both types purpose built factories and facilities helped to attract potential new industry and the new towns do not have the dereliction often found in other parts of development areas. The advantages of new towns are that the population can often live in more pleasant conditions than in the older conurbations and can often find work relatively locally. In general, modern industrial development is likely to be less polluting and industrial areas are unlikely to be too distant from residential areas. This reduces the distance for workers to travel so far to and from their homes.

Whilst the new towns around the London area were originally intended to be self-contained they have not fulfilled this particular need and therefore whilst both Basildon and Milton Keynes did not originally have their own railway stations, they have had new railway stations built. This has meant that commuting became easier.

Since the late 1970s there been a tendency to move away from new towns to the rehabilitation of the Inner-City areas. The new towns have often been fairly effective at segregating different types of traffic, Runcorn with its bus ways and Stevenage with its separate provision for cycles, super buses, etc.

One of the advantages of the new towns is that they could gear their transport infrastructure to what was required now rather than most infrastructure which was developed in the nineteenth century. Purpose built factories could be built to twenty-first century specifications rather than having shapes and sizes, inconvenient for modern industry. Much wider roads and better access to premises could be obtained than in many existing towns and cities.

Cycleways

Whilst the number of cycles sold in the UK is large, little emphasis has been placed on it as a form of transport. Stevenage was one of the first of the new towns developed in the post war period and originally only had a population of 6,000. Currently its population was about 86,000 in 2014. When the town was originally built, it was assumed that people would use bicycles to school and also to work. This is partly because the district is relatively flat. One of the advantages of bicycles compared with cars is that it requires only a small amount of parking space compared with cars and therefore it is much easier to provide access close to factories than to provide for cars. Therefore, for shorter journeys there is a time advantage compared with motoring.

The cycleway system in Stevenage has been mainly geared for journeys to and from school or work although there has also been a limited amount of emphasis upon recreational needs. Where the direct routes would be along primary roads the cycleways have also been made longer using the same land corridor. Generally, the footpaths have been placed between cycle-ways/carriageways so that pedestrians can get onto buses at bus stops. This is better than the pre-war cycle track system where cycle tracks were placed between pedestrian and car tracks which meant that pedestrians then had to cross over the cycleways.

Whilst in the UK, it is often claimed that cycling is only appropriate to countries like Holland which is flat, it should be remembered that the majority of Eastern England is extremely flat.

Enterprise Zones

These were set up by the government in 1981 partly because of the problems of large scale unemployment. They were generally set up in areas of high unemployment including areas which had previously declined, such as the London Docklands area.

They gave three basic advantages to new firms. They gave exemption on rates and taxes on both commercial and industrial property and 100% capital allowance for industrial and commercial buildings, quicker customs facilities and also simplified planning procedures.

London was mainly successful though the Merseyside Development Corporation was somewhat less successful. This is partly because whereas the LDDC was situated alongside the magnet of London this is untrue of the Merseyside Development Corporation, also the Merseyside Development Corporation area is relatively small and split into several sites. Enterprise zones no longer exist.

CHAPTER 19

Measures of Efficiency

How Do Economists Know What Customers Want?

The transport industry often has very fragmented demands. This is particularly true of the bus industry, where often the passengers will not be concerned about the overall efficiency of the transport system, but more whether their bus is likely to be coming along. This particularly applies if they are waiting at a bus stop without a shelter. It is more difficult therefore to carry out market research in trying to find out what passengers want, than it would for an organisation launching a new brand of food. Food organisations could often carry out a pilot survey in one area and assume that the results will be similar elsewhere. The general principles of marketing are important, so if economists found that in one area that a 15 minute frequency produced the best results that they could use the results elsewhere as an indication of potential demand. The transport industry, however, has often been supply rather than demand orientated.

Surveys of the Industry

The railway industry has carried out extensive passenger surveys to try to find out what passengers consider are the most important characteristics, and then try to find out how best it can meet these approaches. One of the main concerns is punctuality.

Different railway operators have had different measures of punctuality. However, even though Railway Focus have found in 2007 are problems in determining how it could be improved to the benefit of the passengers. Passenger punctuality could be more important for an infrequent service than a very frequent service. On the London Underground, most passengers would not know whether their trains in the peak hours are running to time, since they would not know what the timetable is. On the other hand, in the rural areas where there might only be one bus an hour or even less frequently than that, then punctuality would become more important.

Unfortunately, some of the modern technology, such as showing bus services in real time, would not be available where it is most needed (i.e. it will often show this on a reasonably frequent route but not on routes where there are sparse services). However, many passenger and freight operators now use social media to inform customers of any potential delays.

Aims of the Transport Industry

All transport operators would claim that they wish to be efficient. Transport economists would be aware of the nicknames sometimes given to the rail companies (e.g. LNER standing for Always Late Never Early Railway or the M and GN standing for the Muddle and Get Nowhere). Passengers did not always think too highly of the efforts).

Too often, transport operators have been over concerned with the vehicles rather than the needs of its customers as well as its workforce. Economic efficiency can be considered as meeting the demands of the customer with the lowest resource. Whilst the general media as well as often the magazines designed for the industry often have considerable space devoted to the vehicles, the great leaps have often occurred because of the digital revolution.

Computers and Transport

In the transport industry, computers have often been used for accounting purposes. The vast number of calculations that computers can do very quickly makes them extremely helpful in the issuing and checking of invoices to prepare dispatch notes, etc. They are not only used for bookkeeping purposes but also are increasingly used for costing purposes. In the road freight industry, this has been made easier with the tachograph data. It is therefore much easier to allocate repair, maintenance and fuel costs sensibly. It is also possible to use them for scheduling purposes. With road freight operations tachographs disc data can help to determine more effectively when repairs and maintenance can be carried out without losing too much revenue. It is sometimes said that the managers know the workers well so do not need computers. An example of using personal knowledge would be where a driver was always scheduled to be where Manchester United were playing then programming this information into the computer whatever one might think of the football club. Personal information such as when people are taking holidays or need medical appointments will be helpful. Computers are also widely used for stock purposes.

Given the lead-time for deliveries can be combined to reduce stockholding costs. Computers have also been used for payrolls. This has considerable advantages especially where overtime and bonus payments have to be made which is common in the transport industry.

Whilst computers have often been used for calculations, they are now used to transfer data quickly from one purpose to another. Computer programmes in the road freight industry have been developed which will show the distances that have been travelled, the timing of journeys, the amount of down time (i.e. the time which the vehicles are not being used and why this has taken place). This should enable the road haulier to improve vehicle scheduling. This should also enable the haulier to look at the overall life cost of the vehicle and thus to be able to make more effective investment. It can show whether delays were caused at the depot or at the destination, including shops and other retail outlets.

Garbage In Garbage Out (GIGO)

One of the underlying principles of computers as with any other management information system is GIGO. The computer cannot produce information, which is better than that which is put into the system.

Computers Can Help to Reduce Fuel Consumption

Data from tachographs can be fed into the computer and drivers who spend too much of their time accelerating and decelerating can readily be detected. Using computers to reduce fuel consumption is not however confined to road freight. London Transport with the newer Underground stock which was brought into operation in the early 1990s hoped to save fuel through the use of computer controls. Driverless trains are now being used on the Underground as well as on the Docklands Light Railway (DLR). This should help to reduce fuel consumption.

Profitability as Measure of Efficiency

Profitability is the main criterion of efficiency for organisations. However, usually profit has to be compared with something (e.g. very large companies will have bigger profits than smaller organisations so that economists need a profitability ratio in order to compare the firms or organisations). Sometimes, particularly where there have been takeovers or mergers, economists will want to compare efficiency over time and profit ratios will help. Profitability may depend partly upon past decisions about human resources management, including training, and effective recruitment. Immediately after privatisation in 1994, South West trains reduced the number of drivers through redundancy only later to have to recruit others to take their place, would have affected the profitability of the service. Profitability, particularly in the rail industry, depends partly upon subsidies and partly upon the contractual arrangements.

Taxation Factors Affecting Profitability

Subsidies or taxation will affect profitability. In the bus industry, subsidies were given for new buses. Therefore, firms taking advantage of this might have greater profitability, since the newer vehicles might attract more passengers. The rail industry in many countries has been subsidised. The shipbuilding industries have often been subsidised, which means that there have been lower costs for the shipping industry than would occur under pure competition. Road hauliers would usually claim that they have been over taxed, although other commentators such as Pryke and Dodgson in their book 'The rail problem' 1975 would suggest that lorries have not paid their social costs.

German Evidence About Vehicles and Social Costs

Technische Universität Dresden Faculty of Transportation and Traffic Sciences "Friedrich List" Institute of Transport and Economics Chair of Spatial Economics Würzburger Straße 35, 01187 Dresden, Germany in 2013, stated that private vehicles did not pay their full social costs. If in the road system as the EU have stated there is a road pricing system, which reflected the true social costs, then it would be easier to suggest that profitability was a criterion for efficiency. If Concorde was subsidised, then airlines using the supersonic aircraft would have shown higher profits than if the subsidy had not been given. The aviation industry has come under increasing criticism since fuel is not taxed at the same rate as in other parts of the transport industry.

Different Measures of Profitability

There is no one measure of profitability, which can be used to measure efficiency. A large firm may have a large absolute profit compared to a smaller one, but could be much less profitable in terms of profit per person employed or per unit of capital. Therefore, a whole series of ratios could be used, such as looking at the net profitability compared with turnover or gross profitability. Profitability may also depend upon the degree of taxation, which occurs, and sometimes the ways in which firms manipulate the accounting conventions. The Panama revelations 2016 showed how the rich and famous, often manipulated their tax in both capitalist and communist countries.

Flags of convenience ships will usually have lower operating costs, than ones which have their flags in the countries from which they operate. Other shipping operators complain that such ships have a poorer safety record.

Comparisons of Costs

Where firms are operating under similar conditions, the costs per passenger kilometre are one indicator of efficiency. However, it would not be possible to use this where conditions are radically different. A rural bus operator would generally have lower costs per vehicle kilometre than an urban operator has. This is partly because speeds in the rural areas would generally be faster than in the congested town or city conditions. Wages per hour may be lower in some poorer rural areas than in the urban areas where there is a much greater choice of occupations. In air transport, the length of journeys will affect cost per passenger or tonne kilometre since terminal costs will be similar irrespective of sector length. Therefore, any airline operating mainly short haul flights will usually have a higher cost than those on longer distances. This is one of the reasons why prior to the merger when BA was formed from both British European Airlines (BEA) which operated on short haul flights and British Overseas Airline Corporation (BOAC) which operated mainly on long distance flights the two companies had quite different cost structures.

Costs Relating to Quality

The quality of service may also affect costs. A taxi operator will have higher costs per passenger kilometre than a bus or coach operator. On cruise lines, some firms may operate at the top end of the market whereas others have sometimes concentrated mainly on school cruises with much lower levels of accommodation. On the railways too, some operators have given a higher level of service by operating faster trains, whereas others such as former Silver Link have operated a cheaper but efficient service. An operator generally catering for the quality market will have higher costs than those in the mass market.

International Comparisons of Efficiency

It is also very difficult to make international comparisons, especially in an era where currencies fluctuate considerably. This would have been particularly true of the dollar in the period after the British exit from the EU in 2016. Economists therefore would usually use a variety of measures such as utilisation of equipment or staff per unit of output.

Load Factors

This is also a measure of efficiency but not an absolute one. High load factors can be obtained if vehicles are only used in the peak, although this would usually be inefficient for many other reasons. The load factor is defined as:

$$\frac{\text{Traffic actually carried times}}{\text{Capacity}} \times 100\%$$

Capacity

Capacity is not always easy to define, the capacity of an aircraft depends partly upon the seating configuration, the capacity of an Underground train or rush hour train is more difficult to determine. From the passengers' viewpoint, a load factor which is very high might indicate inefficiency, because it probably indicates that they have to stand, and this is one of the complaints which has not only applied to London passengers, but also currently to passengers on the Great Western routes coming into London.

The load factor will depend upon the fares charged. An operator charging low fares is more likely to obtain a high load factor. However, recently airlines have looked at management yield as have the railways and use of the

Internet makes it more feasible to use price discrimination to obtain a variety of fares but also to maintain a high load factor. Charging only one fare or one price makes this much more difficult to obtain.

Staff per Vehicle or Vehicle Kilometre

This is a measure used by some economists as well as many transport operators. However, as a measure of efficiency it will depend partly upon which functions are carried out within the firm. Transport operators carrying out their own maintenance and repairs will employ more staff than those who do not. Since privatisation, most railway operators with the exception of the small Isle of Wight railway are no longer responsible for the maintenance of the track or signalling. Therefore, they will have lower costs and lower staff per vehicle mile than the former British Rail simply because of the lack of vertical integration.

How many staff are required would also depend upon the hours, which the vehicles are expected to run. Transport for London which operates night services as well as a very long day will have more staff per vehicle than a small rural operator, who may only operate buses a few times per day.

Survival Data

One measure of efficiency is survival. For almost all businesses whether in the private or public sector, survival is a major objective. Economists can therefore look at firms in one period and see how many of them survived at a later period.

However, this is not always easy, since sometimes businesses will merge or change names so that it is more difficult to check how far firms have survived. Whilst most businesses in the private transport sector are companies, not all will be and therefore at the very smallest end of the market such as in the taxi trade or road haulage sector, firms may not have been registered as a company which will make it more difficult to judge this factor. There are also some categories where demarcation may make it difficult to judge whether organisations are in the transport sector or not. A typical small shop where the person may make a few deliveries to customers. Do economists classify this as being in the own account sector?

Transport economists observing Great North Eastern Railway (GNER) which operated services from London to Inverness and Edinburgh and Glasgow among others would widely have been seen as successful but its parent company Sea containers ran into financial difficulties and in October 2006 filed for protection against creditors under the US bankruptcy laws.

Customer Satisfaction

Where there is a degree of monopoly power or where large-scale subsidies are involved which exists in much of the public transport sector, economists may need to measure customer satisfaction. There were criticisms of the government in 2016 where the subsidy of £20 million was paid to the owners of Southern rail, even though the company Govia Thameslink (the owner is very profitable). Economists could devise surveys asking passengers whether they believe they are getting value for money and Passenger Focus does this. One of the problems with this approach is that it is not obvious how the passenger can compare this over time. It may also be heavily influenced by the media, which often concentrate on bad news stories.

Surveys can realistically however ask them about the degree of overcrowding and the space allocated to luggage and get meaningful answers.

Economists can ask the operators for information about satisfaction (e.g. the percentage of trains which run late) (e.g. which may be over 5 minutes late at the destination or intermediate stations or the number of times the trains are not of the right formation) (e.g. if there are only 4 coaches when there should be 8). There can also be figures showing the percentages of trains cancelled.

Productivity of Staff

In the freight industry, economists can look at amount per units of labour such as per person hour, but they need as with all comparisons to make sure that they are comparing like with like. The advantage of these comparisons compared with costs is that it gives a reliable figure, which might make international comparisons easier especially where there are violent currency fluctuations which make true comparisons difficult.

Transport economists may wish to have international comparisons. Ports such as Amsterdam, compete with British ports. A modern fully mechanised port such as Thames Haven will obtain far greater productivity than in the more old-fashioned ports where human beings manually do most of the transfer of goods from and between one mode of transport and another.

A port using casual labour might get a higher rate of productivity than one which didn't, but is undesirable for other reasons.

Some firms will do their own maintenance whether it is in the road freight industry or an airline whilst others will wish to have these activities carried out by other firms. Similarly, some firms will wish to carry out their own catering whilst many airlines will wish to get rid of this activity as being a distraction from their core activities.

Turnover per Employee

Some railway companies have used this as a measure of productivity. Again, it will depend partly upon whether or not firms are vertically integrated. If as many people would suggest the railways reverted at some stage to a system of vertical integration (i.e. that they were responsible for the track and maintenance of most of the routes on which they travel), then clearly this figure would tend to be reduced, although it would not necessarily indicate any loss of efficiency.

Training and the Transport Industry

The transport industry except perhaps in the air sector has not been renowned for looking at the number of trained people it has. There is also the danger in the transport industry as in many others through what is sometimes called the accidental managers. This means that people have their own expertise, in whatever field and then will become managers often without any training. In 2007, it was estimated that in UK industry as a whole, that only 1 in 5 managers had any professional qualifications and it is quite likely that this is even lower within the transport industry. Too often the training, which has been given, is relatively narrowly focused (i.e. looking at the immediate operational demands of the transport operator, rather than taking a wider prospective). This is one of the many reasons why the former Chartered Institute of Transport tried in the 1980s to have subjects such as Transport Control and Management which gave people a wider overall view of the transport industry as a whole, since it would be quite possible to know a lot about buses and details of the vehicles without having too much of an idea of how they are affected by both the competition as well as changes in land use planning. This knowledge was often crucial to longer term understanding of the industry and its needs.

CHAPTER 20

Transport Investment

Governments and Transport Investment

Governments, particularly those pursuing Keynesian methods or socialist countries may wish to increase or occasionally decrease the levels of transport investment for macro-economic reasons.

This would be important when unemployment levels were high in the UK as in the 1980s and immediately following the credit crunch 2007 onwards. Lord Keynes, the best-known British economist in his books such as the General Theory 1937 wrote that road building would be one method of relieving unemployment. The multiplier is the ratio between the initial injection into the economy such as the cost of the road building and the total increase in the economy. The multiplier effect may be quite substantial, as the majority of work done may be carried out by relatively low wage indigenous people and therefore little money will leak from the economic system by way of imports or taxation.

The net cost of such measures may well be much smaller than the gross cost if people are otherwise unemployed since the government may otherwise have to pay for social security, it would also gain from employers and employees National Insurance contributions, as well as contributions towards indirect taxation such as value added tax. However, one of the problems involved in such measures is that there is often a considerable time lag between deciding on a major investment and its implementation. There may also be a temptation to announce important investment decisions at the time of by-elections rather than taking a long-term view. If unemployment reduction is the major objective, higher levels of maintenance and repair whether of the road system or of other modes of transport including railways may well be much quicker methods of achieving the same results.

Types of Investment Appraisal Methods

Investment is important for transport firms and there are several different motives for investing capital. Investment relates to the purchasing of capital goods. A capital good is one that is used either directly or indirectly in the production of other goods or services. Examples of these include machinery, plant, tools or equipment. Firms have to replace their capital goods that have become obsolete. This can happen through either advance in technology or just standard wear and tear. Growth also requires investment, as do the refurbishment of buildings such as stations and track in the case of the railways and the purchase of other firms.

The amount of investment required by some transport organisations can be colossal, such as the Channel Tunnel which cost over £10 billion by the time it was finally built in 1994. A proposal to build a railway line between Spain and Morocco would be even more expensive as would be the possibility of having a railway under the Bering Strait between the USA and Russia. The recent development linking the Swiss and Italian rail systems has also been very expensive.

Types of Investment

There are two types of investment: autonomous investment, which is where a firm buys goods to replace existing worn out items, and induced investment, where an organisation buys to expand its operations.

Investment to Increase Capacity

An increase in capacity would be a major reason for investing for example with the West Coast main line improvements which occurred in the early part of the 2000s. It also occurred, when the then British Airports Authority (now Heathrow Airport Holdings) wanted another terminal at Heathrow (i.e. Terminal 5 which has now into operation and again the cost is considerable). The 2017 decision to have another runway at Heathrow would also be very expensive.

Investment to Reduce Costs

An example of cost savings would be as with the use of computers to carry out a number of different activities if the use of cost savings would occur. Accounting is increasingly carried out by the use of computers rather than having vast accounting departments. An increase in revenue would be one whereas with the low-cost budget air carriers; they might well require more computers so that they do not get the problems, which for example Ryanair has incurred.

It is important to have a good idea of what the outcome of each investment option, as well as how this will affect the firm's finances throughout the lifetime of the project. Several methods of appraisal are detailed below. Generally, firms will only invest if the benefits; however, quantified, are greater than costs.

Payback Period

One of the most common is methods of investment appraisal is called the payback period. This is where the firm looks to see how quickly the project or investment pays for itself. For example, if a firm was buying a computer and it estimated that it would cost £10,000 but would save £5,000 per year then the payback period would be 2 years using the formula below:

Cost

Saving per year = payback period

$$\frac{\text{Cost}}{\text{Saving per year}} = \text{payback period}$$

So

$$\frac{10,000}{5,000} = 2 \text{ years}$$

In practice, many small firms use this method, but it has many drawbacks. One of these is that the method does not take into account all the benefits that occur from the investment. In particular, it ignores the longer-term benefits. Consequently, a transport organisation would never decide to send someone on a long-term course such as a degree program since the payback period would be too long. The Channel Tunnel would never have been considered since during the seven years of building there was no revenue and very large costs. In 2017 the French and Italian governments stated that they wanted to have a fast rail link between the two countries. The time after the payback period has been calculated is not taken into account, so this method does not consider the overall profitability of the project in question. An example of this is shown in Table 20.1.

Table 20.1 An example of the payback period method of investment appraisal calculation

	Project 1	Project 2	Project 3
Cost	£10,000	£8,000	£16,000
Return Year 1	£5,000	£1,000	£4,000
Return Year 2	£5,000	£2,000	£4,000
Return Year 3	£5,000	£3,000	£4,000
Return Year 4	£0	£4,000	£4,000
Return Year 5	£0	£5,000	£4,000
Return Year 6	£0	£6,000	£4,000
Payback Period	Year 2	Year 4	Year 4
Total Net Profit*	£5,000	£13,000	£8,000

The Payback Period method of investment appraisal would suggest that Project 1 is the best, as the break-even point is easier to achieve. Further calculation and closer inspection reveals that both Projects 2 and 3 make a greater overall net profit for the firm.

*Net Profit = Total Returns − Total Costs

Average Rate of Return

The Average Rate of Return (ARR, sometimes known as the accounting rate of return) method gives the annual net return on the investment as a percentage of the cost. The formula used is shown below:

$$\text{ARR (\%)} = \frac{\text{Net return per annum}}{\text{Initial cost}}$$

where,

$$\text{**Net profit per annum} = \frac{\text{Net profit}}{\text{Number of years}}$$

It is best explained by referring to the example in Table 20.2.

Table 20.2 An example of ARR calculation

		Project A	Project B	Project C	Project D
Cost		£50,000	£75,000	£90,000	£60,000
Return	Year 1	£11,000	£17,000	£32,000	£7,000
	Year 2	£11,000	£18,000	£30,000	£10,000
	Year 3	£11,000	£18,000	£30,000	£12,000
	Year 4	£11,000	£22,000	£10,000	£16,000
	Year 5	£11,000	£17,000	£8,000	£16,000
	Year 6	£11,000			£16,000
Total		£66,000	£92,000	£110,000	£77,000
Net Profit*		£16,000	£17,000	£20,000	£17,000
Net Profit Per Annum**		£2,667	£3,400	£4,000	£2,833
ARR		=2667/50000	=3400/75000	=4000/90000	=2833/60000
ARR		5.33%	4.53%	4.44%	4.72%

There are four investment projects (A, B, C and D) each with different initial costs, periods and values of returns.

The best project to undertake, according to ARR, is Project A. Whilst this gives a lower amount of net profit than Project C the percentage return is higher. This means that, on average, for each £1 invested, project A returns just over £1.05 in the first year compared to around £1.04 for the first year of project C.

This technique is useful when comparing projects with different starting costs and return periods. It can also be used to compare between the projects and other options, such as putting the capital into a bank account. In the example, if an organisation could obtain a better interest rate at the bank than 5.33% per year then the firm in question may wish to invest the money with the bank.

These first two techniques do not take into account the time effect on money. If the project horizon is over a large number of years, figures should be adjusted to take into account interest rates. These approaches do not do this and so the actual rate of return may be lower and the costs may take longer than is calculated to be paid back in full.

The Effect of Interest Rates (Test Rate or Discount Rate)

Economists wish to translate all the monetary figures to the present day values. As a demonstration, a transport organisation might assume that the rate of discount (*i*) is 10%. They can adjust future values by *dividing* by (1+*i*). The best way to think about this concept is to consider saving accounts. If students have access to a savings account that pays 10% interest per year and someone offers to give you £100 in a years' time or £100 today, then which one would a student chose?

If the person gives the student money now and they put it into the account, it would be worth £110 next year. Therefore, if a student takes the £100 in a year's time then they are £10 worse off than if they had taken the money today. So, if a student is offered £1 in a year's time it will obviously be worth less at the present time, the real question is exactly how much less?

If 'I', denoting investment, is £1,000 then students will find that the discounted amount at 10% is £909.09 one year earlier. Therefore, to obtain £1,000 in one year at an interest rate of 10% they must invest £909.09 now. If they wish to get to the target in 2 years' time then they have to divide i by (1+*i*) again, effectively $(1+i)^2$, in this case 1.21, giving a present value of £826.45. If it is in 3 years' time, it will be $(1+i)^3$ so 1.331 (see Table 20.3).

Table 20.3 An example of present and future values' calculation

Year	2016	2017	2018	2019	2020	2021	2022	2023	2024
Discount	2.14	1.95	1.77	1.61	1.46	1.33	1.21	1.10	1.00
Value	£466.50	£513.10	£564.40	£620.90	£683.00	£751.30	£826.40	£909.00	£1,000

Discounted Cash Flow

Discounted Cash Flow (DCF) considers this important point. The basic logic of discounted cash flow is that returns in future years do not have the same value as cash now. In most cases people or society, prefer to have money now rather than money later. This is sometimes called 'social time preference'. This is conceptually different from the effects of inflation although the ideas overlap.

Interest rates are sometimes denoted by the letter *i* and occasionally by *r*.

Economists can outline the two concepts that are called Net Present Value (NPV) and Internal Rate of Return (IRR). Most people are used to the idea of going forward (i.e. if they put £100 into the bank now it will be worth say £105 in a year's time). They do this calculation by multiplying by (1+i). If i is 5% they multiply the original amount by 1.05. If i is 10% they *multiply* the amount by (1+i), which is 1.10. If it is 15% they multiply it by 1.15, etc.

If they put the money aside for a second year in the bank they multiply it by (1+i) again or simply (1+i)². If they put the money aside for a third year they multiply it by (1+i)³.

The interest rate that is to be used will depend upon the circumstances surrounding the decision. For example, if we can borrow from the bank at 20%, then the test rate or discount rate needs to exceed 20%; otherwise, we will not be making any profit when financing the project through bank loans. Organisations wish the rate of return to be more than 20% in this case because in most investments there is an element of uncertainty. This higher value allows organisations to take into account some of the possible problems and any over estimations of the returns.

If an investor put £100 into a bank for a year, then they are saying that we prefer the money in one year's time to £100 now. If we do not do so, then we are saying that we prefer the money now to later. When firms invest in machinery or capital they are making these decisions about preferring money now or later. One of the problems in practice is that there is always an element of uncertainty with investment projects. In the case of the Channel Tunnel, about which there has been a great deal of controversy, there is no certainty that the £10 billion that has been spent on the construction will be recovered.

Ensuring that Benefits Are Greater than Costs

The aim, generally, is that the total benefits are greater than the total costs.

In this case, the organisation will wish to ensure that Total Costs 'C' in all years up to the final year (n) is less than Total Benefit 'B' gained in those n years, adjusted by the discount factor i. Therefore, we want:

$$C_0 + (C_1/(1+i)^1) + (C_2/(1+i)^2) + (C_n/(1+i)^n) < B_0 + (B_1/(1+i)^1) + (B_2/(1+i)^2) + (B_n/(1+i)^n)$$

This can be reduced to:

$$\Sigma(C_n/(1+i)^n) < \Sigma(B_n/(1+i)^n).$$

A demonstration of this formula can be seen in the following example (see Table 20.4).

Example: A firm wishes to update its distribution vehicles and has been persuaded by an automobile manufacturer that investing in one of its new vans, which costs £20,000, will produce savings of £4,500 each year for 5 years. The company does not have the capital to invest in the van now so they plan to get finance from the manufacturer at a rate of 6% per annum. What is the total discounted cash flow for the next 5 years? Does this seem like a good option to pursue?

Solution: The organisation is spending £20,000 now, say Year 0, and will receive £4,500 in Years 1, 2, 3, 4 and 5. The interest rate is 6%. The table below

Table 20.4 An example of calculating benefits

Interest rate	6%		
Cost	£20,000		
Benefit per year	£4,500		
Year	Cash flow	Formula	Discounted Cash flow
1	4,500	$4500/(1+0.06)^1$	£4,245.28
2	4,500	$4500/(1+0.06)^2$	£4,004.98
3	4,500	$4500/(1+0.06)^3$	£3,778.29
4	4,500	$4500/(1+0.06)^4$	£3,564.42
5	4,500	$4500/(1+0.06)^5$	£3,362.66
Total	£22,500		£18,955.64

shows the cash flows and the discounted cash flows using the formula above.

Total Discounted Cash Flow is £18,955.64
Here $\Sigma(C_n/(1+i)^n) > \Sigma(B_n/(1+i)^n)$
Or more simply £20,000.00 > £18,955.64

As the total discounted cash flow value is less than the total costs of the project it should not be pursued. This would not have shown up had the discounted cash flow method not been used.

Net Present Value

A useful term in investment appraisal is Net Present Value (NPV), which is the value of investment income minus the cost.

NPV = Present value of returns − Costs
If NPV = 0 then the project should breakeven.
NPV > 0 then the project should make money
NPV < 0 then the project should lose money
In the example above NPV = £18,955.64 − £20,000.00 = −£1,044.36

Therefore, the project is likely to lose money.

The Rate of Return

The mathematical formula to work out the rate of return is quite complicated. We can however demonstrate by working backwards with a 2 year example. If we are offered £121 at the end of a 2 year period for a sum of £100 now, then the rate of return is 10%.

This is because $100 = \dfrac{121}{(1+i)^2}$ which rearranges to:

$$100(1+i)^2 = 121$$

$$\sqrt{100}(1+i) = \sqrt{121}$$

$$10(1+i) = 11$$

$$10 + 10i = 11$$

$$10i = 1$$

$$i = 0.1 = 10\%$$

The Internal Rate of Return (IRR) calculates the percentage return on a project at which Net Present Value equals 0. If the market rate of interest is known, then any project that provides an IRR above this value is calculated to be worthy of investment. The main advantage of using this method is that it can be used to compare various different projects more fairly than the other techniques discussed in the chapter. The problem to this technique is that it gets increasingly difficult to calculate the IRR of projects lasting more than 2 years. This is because using a square root is not too difficult to deal with, as shown above, but cube roots (used for 3 years) and above tends to be more difficult to handle without computer software.

Types of Benefits

Generally, organisations can subdivide benefits received by investing into 3 categories, although these may overlap. They are:

Cost Savings

For example, installing a computer could save money by reducing the administrative tasks carried out by staff.

Increases in Capacity

The computer will allow the organisation to do more things such as store larger quantities of information.

Increases in Revenue

A more powerful computer may allow the organisation to serve more customers.

Often economists are not certain what benefits will result from the investment. It is generally easier to ascertain savings in costs rather than increases in revenue.

Sensitivity Analysis

Because it is often so difficult to estimate the cost, and even more difficult to quantify the benefits, transport organisations may wish to carry out a form of sensitivity analysis. This means that they test how sensitive the project is to possible changes whether in interest rates, demand or initial costs, etc. If the project is very sensitive to changes, we they may carry out further work to see if they can make their forecast more accurate in any way. However, if the project were robust, meaning that the range of possibilities does not greatly affect it, we would proceed.

Problems of Forecasting

All the investment appraisal techniques in this chapter are based on predicted values based on current assumptions. The return on an investment may be overestimated or the cost underestimated. The interest and inflationary rates may change over the period of the investment and so affect its outcome. These factors mean that there is a need to build in some form of margin of error. Nobody knows exactly how the economy may change in the future and so managers must not blindly follow the maths without considering the external influences upon those decisions and their possible changes.

Long-term forecasting is extremely difficult. Economists trying to estimate the traffic for a new road bridge, once the project was under way could use regression and extrapolation methods. If, however, they were trying to estimate the traffic in 40 years' time so many changes could take place that it would be very uncertain. Any number of factors would affect the usage of the bridge including changes in holiday patterns, growth or decline of different types of transport, business communications patterns, population changes or even natural disasters.

Independent and Interdependent Projects

Sometimes, the portfolio of project options may be independent. Sometimes the projects may be interdependent. An independent project is one that, if chosen, has no influence on the outcome of any other project. Interdependent projects are those that affect the outcome or otherwise influence each other.

When they are interdependent, it is worthwhile considering Project A alone, Project B alone and 'A' and 'B'. If a firm were to invest £40,000 in a new computer system for its head office, it may expect to make savings of £4,000 a year. If the firm kept its existing system but trained its workers to use it more effectively, this may cost £8,000 and give savings of say, £1,000 per annum. If, however, the firm invested in the new system and training then the figures could be different.

In this case, the benefits of the project as a whole may be greater than the sum of its parts. The total cost may be lower as a deal with the supplier, may be possible. The returns may also be higher than the two individual projects combined. In this case, it may mean that the combined costs are only £45,000 saving £5,500 per annum.

The reverse may also be true. If an organisation wishes improve their existing paper filing system and to install a computer system, then either may be worthwhile on their own but not necessarily together.

CHAPTER 21
Economies of Scale

Economies of scale can be defined as the advantages of being big. This can apply in the transport sector to the vehicle, the depot or the terminal, the size of fleet, the organisation or the way. There are also diseconomies of scale which can be defined in a similar way.

Economies of Scale (Vehicles)

Larger vehicles may be proportionally cheaper to buy. This is partly based on the arithmetic rule that volume rises faster than the surface area so that there is proportionately less material used in the construction of the larger vehicle. Where constraints are not imposed by the way or by the terminals, there can be very large vehicles such as the very large crude carriers (VLCC) at sea or even the European airbus and the Boeing Dreamliner.

Crew size does not rise proportionally. One driver is necessary for the largest or the smallest road haulage vehicle, which is currently allowed in the UK or most European countries, typically around 44 tonnes. Whilst larger ships require more than one member of the crew a 500,000 ton tanker does not take 10 times the number of the 50,000 ton vessel. Similarly, a jumbo jet with 500 people does not require 10 times the number of crew of a 50 seater jet.

Fuel consumption is also proportionally better with larger vehicles partly because the crew themselves with their weight are using some fuel. The space that they occupy will be proportionately smaller in larger vehicles.

The Wood Report 1981

This was set up in 1981 to look at the socio-economic and the environmental effects of banning heavy lorries within London. The Wood Report had a variety of independent members including people from the Freight Transport Association, Multiple Retailers Association.

The report looked at the volume of traffic in terms of vehicle kilometres by laden weight as well as vehicle hours as well by laden weight as well as the origins and destinations of the traffic and also looked at average trip length and drop size.

The report looked at the arguments for and against lorry bans and suggested that the witnesses who thought their complaints about lorries were only from unrepresentative environmentalists were seriously wrong. They noted that the then Greater London Council (GLC) had powers to impose bans and it looked at the effects of the Windsor Cordon ban. One of the suggestions was that they might look at the possibility of paying fees by having an axle number which was readily identifiable.

Perhaps the most important concept was looking at the variety of types of bans which could be imposed within the maximum weight limits of 7.5, 16 and 24 tonnes and also by the types of road network.

The report pointed out that sometimes that banning a 30 lorry would not necessarily mean having just over two 16 tonners to replace it, since there would be scope for breaking bulk and transhipment and some of the larger lorries would have not used to their full capacity perhaps being utilised in off-peak times or possibly being towards the end of their journeys.

The report also looked at the possibility of transferring from road to rail and showed that currently that rail was a small percentage of total tonnage but carried a fairly large amount of goods which otherwise might have been carried by the heavier lorries. The report also looked at the possibility of exempting railhead traffic which would boost the railways share of the market but would reduce the environmental gains.

Limitations on Size of Road Vehicles

Limitations on size may be imposed by the way, the terminal, the loading or unloading times or government or other legislation. All European and nearly all countries will have construction and use regulations which impose maximum limits on the amounts of weight that can be taken on the roads by lorries and there are also similar regulations for the number of passengers that can be carried on buses or coaches. Sometimes vehicles over a certain size such as lorries may be prohibited at certain times of day. From the government's viewpoint, it is not always the maximum weight that is important but the weight imposed upon the roads by the axles. This varies approximately according to the fourth power of the axle so that if there is a doubling of the weight on the axle with vehicles being overloaded; the wear and tear on the roads is likely to be 16 times as great. There may also be height restrictions on individual roads imposed by bridges and in some cases maximum height restrictions imposed by the government. There are nearly always width restrictions which are typically about 2.5 metres except for abnormal loads where special regulations usually apply. These width restrictions also apply to buses and coaches.

There have not always been limits upon heights on a national basis apart from those posed by bridges or tunnels, however, apart from double-decker buses the constraining heights are usually those imposed by the problems of loading and unloading the goods or passengers is a major constraint on height limits. Whilst lengths have been imposed mainly because of manoeuvrability; this is less likely to be true in a sparsely populated country such as Australia where large lorry trains can be found, since traffic is light and there are obvious economies of scale on long journeys through difficult territory.

Limitations on Size of Rail Vehicles

The rail services have width limits and the UK has a smaller loading gauge generally than many European countries apart from on the older Great Western Routes many of which were built by Isambard Kingdom Brunel to the 7 ft. gauge, even though this was abandoned in 1892. The widths are still often there so that it allows faster running. The last UK mainline railway to be built, the Great Central line which started to operate out of Marylebone station in 1899 was built with the possibility of a channel tunnel in mind. There have been suggestions that there could be a reinstatement of the disused part of the line to allow through freight traffic on from the continent via the Channel Tunnel.

There are not usually any restrictions per se on the length of trains and this was shown in the Second World War in the UK when often the railways could operate trains with more than 20 passenger carriages to avoid the problems of manpower shortages.

In the USA where there are many long routes, the freight trains can be over one mile long.

Size of Ships

There are no restraints on the size of ships generally except on the inland waterways and canals. The 1956 Suez crisis meant ships could not use the Suez Canal. Therefore, many used the Panama Canal instead. The term Capesize ships has recently come into fashion to denote those which are too big to go via either the Suez or Panama canals and so to have to go via either Cape Horn or the Cape of Good Hope. The Panama Canal has been widened. Many UK canals have 7 ft. locks which means that the width of the boats is limited. The length is limited by the locks which are often 60 ft. Therefore, narrow boats sometimes called long boats were around 60 ft. long to gain the maximum width and length which was possible. The broad canals were limited by the 14 ft. locks.

As ships have become larger the draught (i.e. the depth of water that they require has often become larger). This can sometimes be overcome by dredging as in Poole Harbour on England's south coast where dredging means that the ferries can coexist with the many pleasure craft which need very little depth of water.

In shipping the way is free but the depth of water is not always sufficient for the very large modern vessels such as the Super Tankers (the technical term is very large crude carriers, often abbreviated to VLCC). Some of these therefore will moor near Bantry Bay in the Republic of Ireland leaving smaller vessels to serve other ports. In particular, the Suez Canal cannot take the very largest ships and many of these will have to go via the Cape Route, even though the Suez Canal has been enlarged. The same has also been true of the Panama Canal where the Panama Canal has also been enlarged to take larger vessels.

In shipping the BACAT and LASH systems allow one barge to be detached from others in about 15 minutes which can then be loaded or unloaded separately without all the cargo having to be held up in a port and without valuable capital (i.e. the ship having to spend more time for relatively small loads). Ships have reduced loading and unloading time partly because of the digital revolution and also straddle carriers. Whereas in the 1950s ships even on the longest distance routes including those from the UK to New Zealand would spend half their time in ports because of the time unloading and loading. Currently ships spend far less of their time in port and so fewer ships are needed for the same volume of traffic. The use of roll on roll off ferries for short sea routes has also increased efficiency.

Size of Aircraft

There are no restraints on the size of aircraft apart from the constraints imposed by the terminals. Airships could be about 800 ft. long (around 240 metres).

Limitations on Size of Terminals

There are several different ways of judging the size of railway stations which include the number of platforms or the number of passengers using the station during a specified period. Railway stations can have a very large number of platforms as is shown by Waterloo, Victoria, London Bridge and Liverpool Street just in the London area alone.

Generally, termini will have a larger number of platforms than stations such as Birmingham New Street which mainly caters for through trains. This is because the stations which are termini will require time for the locomotives or multiple units to turn around. Sometimes time is also allowed for cleaning so that the train is taking up a platform whilst this is being carried out. In the UK, Waterloo had the largest number of platforms (24) until the Eurostar services started in 1994 and so the station was extended. Economists will observe that often the number of platforms has depended upon past history rather than operators deciding on an optimum size. The larger number of platforms can be helpful for passengers since they may have a wide variety of destinations and this avoids having to change stations.

However, sometimes going from one part of the station to another can be difficult if there are many steps. In the UK, there are more luggage carriers which can take passengers and luggage from one part of the station to another.

Larger railway stations can have more facilities.

London Bridge station has been changed considerably to give better access and faster times for passengers between different platforms. Birmingham New Street has been dramatically changed before its reopening in 2016 with better access to the local tram routes.

Problems of Termini in Major Towns and Cities

Whilst the London railway termini are there mainly because of historical environmental objections to through running and to stations within the City of London. It could be argued that having termini in major cities such as London is suboptimal since the land used for trains waiting in platforms is unhelpful.

The land is not being used optimally especially given the high price of land in cities such as London. At one stage, former British Rail wanted to have trains running through Victoria to Kings Cross which was one of major patterns of travel partly for this reason.

Both Crossrail now called the Elizabeth line and Thameslink do not have this disadvantage, also since the location of the London termini is often not where passengers would wish to travel to and they have to make other onward journeys by either bus or Underground in London a through service would be more helpful to many passengers.

Loading Gauge and Effect on International Traffic

The loading gauge to Felixstowe on the East Coast has been improved so that the port can take the larger container traffic on its routes.

The Channel Tunnel which was finally built in 1994 was criticised for being over budget. Some economists state that it could have been built to accommodate piggy back trains or the newer 9 ft. 6 high containers. Piggyback trains are where the lorries can travel on the train with the consignments. This has the advantage of making loading and unloading easier and also that one driver can make the whole journey but the disadvantage that the lorry is not being used. Currently the smaller loading gauge means that British rolling stock can usually go through to the continent, but the converse is not true.

Railway Depots

Historically the railways had many private sidings often for the benefit of small farmers. Milk trains were a fairly common feature until the 1960s. However, the disadvantages of small sidings were that an infrequent service was often run which was very expensive and the Beeching report illustrated this. Therefore, much more attention was paid to concentration depots. With fewer depots there could be a more frequent service and therefore overall times would be improved.

Larger Warehouses

Warehouses have often been located near the intersections of motorways. They have become larger over time, partly because of the high cost of some equipment especially if the goods are being transferred from one mode of transport to another.

Loading and Unloading Time and Effects

Loading and unloading time may also impose limitations upon vehicle size. In the London area, bus tickets were more expensive if they were issued on the bus than either via the Oyster card or via machines at bus stops. Now all bus tickets are issued off the bus. In most continental countries, the majority of the revenue is now obtained off the bus.

Many countries have carnets of tickets, which can be sold at local newsagents and thus few people have to tender fares on the bus.

Size of Bus vs. Frequency

Recently there has been more interest in mini buses, which are now used very extensively in the Exeter area because the high frequency of a service is part of the attraction to customers. On the other hand, where there are already very frequent services as in the London area there has been greater use of articulated vehicles, which can carry large numbers of people. London unlike many other European cities has always had double-decker buses. They can be helpful since they hold more people without the problems of manoeuvring which articulated buses have.

Road Freight

The increasing use of containers or demountable bodies means that utilisation of larger lorries has improved and one of the main reasons why the road haulage industry pressed the Armitage enquiry in the early 1980s for larger lorries. In 1983, the larger lorries (38 tons) were finally allowed instead of the previous 32 tons. Larger vehicles, however, often mean a less frequent service and this explains why there are a wide variety of different sized vehicles operating at any one time. Currently the road lobby is putting forward the idea of a maximum 60 tonne limit.

Railways

In the UK, since privatisation in 1994 the number of passengers has doubled. There has been concern about increasing congestion both at London and in the Birmingham area. One possibility, which has been discussed in the UK is the use of double-decker trains, which have been used in the Netherlands. On the other hand, some economists state that it would be much better to have longer trains even though this would mean that platforms have to be extended. Part of the problem with privatisation is that too little concern has been given on routes such as those through passing Birmingham about trains which use congested track, but which themselves are nowhere near the maximum length which could be used. One possibility which would be unpopular with passengers, who prefer through services, might be to only allow larger trains in the peak hours (i.e. longer ones into the London Terminals leaving passengers to change trains at the more outlying stations). This might be a cheaper short-term solution to the problems than expanding platforms or altering "The Way".

Aviation

More attention has been paid to quick turnaround time of aircraft at airports especially with the advent of the budget carriers such as Ryanair which manage to keep fares low partly thorough the minimising the unloading and landing of the aircraft. Any time taken with the aircraft standing at airports means that the firm will not have their main asset earning money. For an entertaining account of the way that Ryanair grew to be one of the giants in the aviation business the reader is recommended to read Siobhan Creaton's book on Ryanair.

Research on Size of Depots, Garages, etc.

Whist in other industries there has been a considerable amount of studies at the plant level (i.e. factory) noticeably by Prof. Bain. This has been less true in transport, although The Chartered Institute of Logistics and Transport has taken a great interest in logistics. Generally, with the development of unit loads and the high cost of handling equipment, there has been a concentration of freight on far fewer depots and the number of marshalling yards has been drastically reduced. The marshalling yards are where shunting operations took place so that individual wagons could go from one train to another. There was considerable duplication of marshalling yards in the Carlisle area where many different rail routes converged and so a larger one would have had the advantage of giving a more frequent service to many rail freight customers. The Beeching report did rightly highlight the slow overall times for many customers.

On the other hand, perhaps the Beeching Report took too little interest in the "door to door" traffic which is possible for example in the car industry where Fords ironically often used the trains to carry fright from Lancashire to their plant in Essex. The Freightliner service has been one of the successes from the 1960s. At the depot level, there can also be economies of scale since expensive equipment will only be helpful if it receives sufficient utilisation. The limitations of accessibility may, however, impose an upper limit on the size of depots and therefore economists would expect the optimum size of a bus garage to be lower in a rural rather than an urban area. On the railways, however, there has been a reduction in the number of depots where maintenance can take place and sometimes this has led to criticism about the poor utilisation of stock. The concentration of depots on railways services may enable a more frequent service than the former local station service where overall journey times were often very lengthy. At the depot level, problems of size may occur with management. Whilst a larger depot may gain from the specialisation of labour, there can be problems of communication and perhaps an impersonal atmosphere with larger depots. The standard problems of too lengthy a chain of command, problems of vertical communication partly upward and lateral communication apply within the transport industry.

Economies of Scale of the Fleet

Bus Fleets

The size of the fleet economies of scale will depend upon the size of orders placed. Clearly London Transport which ordered Routemasters would have gained these since they were in operation from 1956 when they replaced the RTs until their last days of normal operation in 2005. Having a fleet of the same types of vehicles has many advantages. The owner will gain since spare parts can be ordered in large numbers and also maintenance staff will know the vehicles better than if they have to deal with large number of different types of vehicles. Drivers know the vehicle well.

Road Haulage Fleets

Whilst there are the usual economies of scale of bulk purchase, there is little evidence that generally larger fleets have great advantages over smaller ones, the analysis gets more complicated, since many firms have their own account fleet (i.e. using vehicles which are used to carry the firm's goods rather than to carry goods for other customers).

Shipping

Historically ships were ordered on a one-off basis partly because the cost of many of the larger ships was such that few firms were going to buy more than one at a time. The QE2 launched in 1967 cost £36.5 million. There were some exceptions such as the Liberty ships in 1944 where the then wartime coalition government wished to have ships built as quickly as possible. Given the life of the ships this led to an accelerator effect on the shipbuilding industry in the late 1960s and Austen and Pickersgill in Sunderland, North East England produced the SD 14 which was a standard sized design ship built partially to replace these liberty ships.

Economies of Scale of the Firm

It is often difficult to compare economies of scale looking at different transport firms. This is partly because even in the same mode of transport they may operate in different sectors. In the bus industry, larger companies in the UK have mainly been engaged on stage services (i.e. they call at many bus stops). Smaller companies are more common in contract hire work (e.g. taking passengers to and from schools and colleges or firms in the morning or evening peaks). Sometimes, they may also operate excursions and day trips where the more personal service of a small firm is helpful. Within the bus industry, there have been differences between the rural routes which are often more suitable for firms which sometimes could use the vehicles for dual purposes (i.e. perhaps running a mini bus in the morning and even using it for a parcels service during the day). On the railways since privatisation in 1993 there have been many attempts to look at the productivity of the different operators. However, there are major differences between the operators. On the Isle of Wight, there is a shuttle service with very old London Underground stock between Ryde and Shanklin in contrast to services such as the successful Great North Eastern Railway (GNER) which operated between London and Edinburgh.

In air transport, especially when the IATA fares structure applied, larger organisations often concentrated mainly on the scheduled services, whilst smaller ones predominated in the chartered flight sector. This was complicated even more since often tour operators operated their own aircraft and therefore they were operating services, not merely for ordinary passenger purposes but as part of the package holiday market.

One hypothesis is that economies of scale are more likely to exist in liner traffic (i.e. on regular routes operating to particular destinations on a fixed time table, than in tramp type services which operate according to demand). Economists will find it difficult to find appropriate data. The newer budget airlines often operate from the smaller airports, which have lower costs.

In the road freight industry, the parcels sector has been dominated by a few large firms such as FedEx. Larger organisations are well-known. Potential customers sending small volumes of goods will not wish to spend much time on search costs.

In the road freight industry, there is relatively little evidence of economies of scale. There are still a large number of small operators including owner-drivers as well as the large-scale firms such as the Transport Development Group. The National Freight Consortia (NFC), formerly the National Freight Corporation, was de-nationalised under the 1980 Transport Act. The NFC believed in economies of scale since in the mid-1980s it took over SPD (the transport branch of Unilevers).

What Factors Might Lead to Economies of Scale for Transport Operators?

Operators may gain from horizontal integration. This means the firms expand either through organic growth (i.e. on their own or possibly though takeovers and mergers at the same level). Transport economists can observe the 1923 merger into four main UK rail companies, when the government after the experience of the First World War showed that large companies gained from economies of scale.

On the other hand, firms often particularly in the rail industry used vertical integration so that they were responsible for their own building of locomotive and rolling stock. Many of the rail managers were engineers and this led them to believe that they knew what was best for their particular companies. Even after nationalisation they continued to build their own locomotives, etc. and often this continued within British Rail Engineering Limited.

The maintenance was also often carried out at depots such as Swindon, etc. which had become major railway towns because of the engineering work.

There were, however, complaints that the railways after nationalisation did not gain enough from such economies of scale. They did not consider other parts of the rail network and the Western region had many different types of wagons with incompatible coupling.

Complaints about lack of vertical integration have continued after privatisation with comments that the power supply for the third rail system in the South East of England is inadequate for the newer more powerful trains including the Eurostars which use more electricity than before. This problem should partially cease now that the Eurostars have their own separate track into St. Pancras.

Categories of Airports

There are many ways of categorising airports. One category is hub airports where people can change from one aircraft to another. There are also regional airports. The smallest airports sometimes called airfields are where people may be able to fly themselves or alternatively have private jets and other aircrafts.

Airfields often reduce accessibility time. Hub airports have disadvantages. Space near such airports is often restricted, delays on one flight can have effects on others and the problems for people changing aircraft can be considerable. If there are hold-ups whether it be a strike by air traffic controllers or simply problems with equipment the delays can be very considerable. If the terrorism threat is important then hub airports are more vulnerable than more scattered ones.

Contrast of Airports and Shipping Terminals

Airports have almost grown into mini towns whereas many shipping passenger terminals such as Harwich or Dover have few facilities. The time taken to travel within a major airport such as London can be considerable and even though travellators may be used some of the advantages of the quicker flights can well be offset.

The Terminal 5 at Heathrow has 127 new shops but only 700 seats for passengers. Shopping and retail outlets accounted for a very large amount of the revenue earned by BAA.

Prices, however, in the shops can often compare favourably with those in the high street and passengers often have dead time which cannot be used for other purposes, since the waiting time at airports has become longer.

How Can Economists Determine Whether Economies of Scale Exist?

Profitability

This is an obvious measure, but a large profit could indicate monopoly power rather than efficiency. Until the time of privatisation in the 1980s and 1990s, there were many nationalised industries in the public sector. Sometimes lower profits resulted from the railways having to pursue social rather than profit maximising policies. This would have applied on the railways, which could have been more profitable if they had axed many rural routes, which would not have been financially viable. Also in the early 1970s the government often held down prices because of prices and income policies in an attempt to hold down inflation.

Even within the privatised sector for a while prices were held down by a formula of inflation minus a fixed percentage, as there were fears of monopoly power. However, in 2007 the government allowed increases above the rates of inflation on certain regulated fares, and in other cases for unregulated fares the increases were even greater.

The profitability shown by accountants depends partly upon capital gearing. A transport organisation which is financed entirely by loan capital will show a lower profit than one that is financed almost entirely by shares. This is very important in the case of the Channel Tunnel, which cost over £10 billion to construct. Clearly if it had been financed entirely by shares, then its losses would not have been so great since it manages to cover its operating costs but not to repay the amount of capital put in. The capital gearing also makes it difficult to compare public and private sectors since much of the public-sector capital has had an imputed rate of interest attached to it. Very small firms such as the owner operators, particularly in the road freight industry and sometimes in the bus industry, may overstate their profits (from the economist' point of view) since a large part of the profits may reflect the opportunity cost of wages, which the proprietor could obtain outside. This would also be true of taxi operators. This makes it therefore difficult to use data such as that used by C.D. Foster in the road haulage industry in the early 1970s. Profits may also fluctuate considerably according to whether there is a boom or recession. Clearly in the boom time profits will be higher than in a recession and again the different capital gearing will tend to make some profits seem much higher than others, even when both firms are operating with the same degree of efficiency.

For the public sector in particular, profitability is not necessarily the only criterion since many economists state that we need to take much more account of social costs. In the absence of a road pricing system, which took into account these costs and benefits, it is difficult to compare efficiency. In the road haulage industry, in particular there are the TWO main sectors a) own account (i.e. firms who send their own transport, their own goods mainly through operating their own fleet, though occasionally they may hire in vehicles). Sometimes as with Tate and Lyle they may also use their own vehicles in order to try to obtain return loads, which they did not do in the past. There are also b) the hire and reward firms who cater for other people.

Small firms as the Bolton committee suggested in 1971 may have independence as a major objective. Even in larger firms, proprietors may not wish their firms to grow too much if independence was to be sacrificed as a result of expansion.

Profitability is not necessarily a reflection of current management ability, since the effects of past management may be important. Many of the railways managers in the early 2000s often complained about lack of investment under the former nationalised regime. Profitability is particularly likely to be affected where the infrastructure is not good enough.

Limitations on Economies of Scale

The size of the market may limit the size of the firm. Many taxi firms and operators in rural areas may remain small for this reason. Clearly a taxi service operating on a remote island may have quite satisfactory profits but may find it difficult to expand. The same may be true of the owner driver in road haulage.

There may be limits of economies of scale because of the management, may suffer from the typical problems off too lengthy a chain of command or too large a span of control. Sometimes the nationalised industries were usually governed by the ultra vires clause, which prevented the industries carrying out logical diversification. Prior to nationalisation the rail companies were often large-scale bus road haulage operators and in one case (GWR) also operated an air service. They had diversified so that they could give a complete service to their customers. The railway companies also ran their own hotels, had their own catering staff and in some cases also ran shipping companies. Sometimes this was to compete in different ways as with the Tilbury – Gravesend Ferry which enabled operators North of the Thames to compete with those South of the Thames. Sometimes in the short run there may also be problems of finance or land availability. In particular, within the aviation industry the number of slots which is available at Heathrow maybe limited and this in turn will impose limits on the size of the firm at least in the short run. In the case of the rail operators a new firm The Wrexham Shropshire Marylebone Railway Company (WSMR) which wished to operate from North Wales to London via Shrewsbury was prevented from doing so since it will compete with Virgin, which has a franchise, which excludes this possibility but finally won permission to run 5 trains per days via Gobowen.

Diseconomies of Scale

One of the problems with the rail industry as well as other transport modes is that there may be conflicts between the different functions. The problems of jealousy arise when firms have been merged or taken over. It was often rumoured that the network which the railway ran down or improved were on the basis of past loyalties rather then looking at the network as a whole to see which was best.

CHAPTER 22
Problems of the Peak

The problems of the peak in transport refers to the tendency in transport for demand to fluctuate with both high points and low points, sometimes occurring on a daily basis and sometimes weekly and seasonally. Sometimes the transport operator may be fortunate and may have a peak in one area or time, which may fit in with a low point in another area or time. Thus, for example in aviation the peak movements for passengers at weekends which may fit in quite well with declining freight movement at that time; whilst low passenger movements at mid-week fit in well with peak movements of freight. This is why it is still largely true that the best place for freight traffic by air is on the passenger plane at a time when there are empty seats in the cabins.

Daily Peaks in Road and Rail Travel

The peak arises on the passenger side, mainly because of the timing of the working day. In the UK for example, office hours usually start between 8:30am and 9:30am and end between 5:00pm and 6:00pm in the evening. Therefore, since transport is a derived demand, it has a peak for arrivals between 8:00am and 9:30am and with people starting their journeys before this, and similarly starting their journeys in the evening from about 4:30pm to 6:30pm. Factories tend to have similar peaks arising at the end of shifts, but proportionately more people tend to travel by car, thus the peak causes problems on the roads rather for the transport operators. It also tends to cause slight problems of the peak for freight transport. It is not only the journey to and from work that causes peaks, since educational travel is often quite substantial. This has become even truer since many smaller schools have been closed. It is also truer, since many people will send their children to schools at quite a substantial distance. This sometimes makes the peak worse especially in the morning. Whereas in the past children often walked to and from primary schools this has become less true. In some areas, schools have adopted the "walking bus" principle whereby one adult is at the front of the group and another at the back. This has a number of advantages (i.e. that children have the advantage of socialising and there is less risk of accidents apart from easing the problems of congestion at these times).

In some places where people live near their work, there may also be a mid-day peak. There may also be a peak on a weekly basis, for example on Friday evenings particularly in the summer months when people take journeys for leisure purposes associated with the approaching weekend.

Seasonal Peaks

There may also be seasonal peaks for example the short sea ferry journeys from England to France have a great deal of passenger traffic in July and August, but comparatively little in February and March. The same is true to a lesser extent of the Channel Tunnel. At one stage where there were both chartered flights for holidays as well as scheduled flights a great deal more traffic occurred in the summer months than in the winter. This peak concept is still true and so the low budget carriers will often give much cheaper rates at some times during the year as will other air transport operators. There are also seasonal peaks in freight, including in most countries festival periods. One is at Christmas as people send each other presents. The wine and spirits trade does considerably more business towards Christmas time. Perhaps more surprisingly the computer trade is one in which it has been estimated that perhaps three quarters of personal computers are brought towards Christmas time and therefore in turn this will accentuate the peak for freight deliveries.

Rural Area Peaks

In rural areas, the freight peak will arise with the transport of agricultural products, although sometimes with refrigeration, peak traffic reduces to a steadier pattern with surplus being carried in a frozen form, at a time, which is more convenient for the transport operator. Near ports historically there have often been peaks especially in the morning as fresh fish has been transported to other towns, though again freezing the products has offset this.

The Problems Associated with Peaks in Transport

There are two main problems associated with peaks in transport. They are a) the poor utilisation of assets; and b) staffing problems. Economists can look at each of those in turn.

There are four classes of assets in transport, the way, the unit of carriage, the unit of proportion and the terminals. Peaks in transport require the operators to provide the assets in such quantities that peak traffic can be accommodated economically, although they will inevitably be a trade-off between providing extra facilities and greater congestion. In general, there is an inclination of those who provide facilities to delay extra provision until congestion begins to become more unbearable. This can readily be seen with the increasing number of complaints as rail traffic has increased in years, since privatisation in the UK in the 1990s and passengers, particularly on some of the Great Western routes, have complained bitterly about lack of seating capacity in 2017. The provision of facilities is therefore to some extent piecemeal and improvements are made where they are most necessary rather than being clearly thought out.

Privatisation of the railways may have made the matters worse rather than better, since the franchises have been relatively short in most cases, ranging typically from seven years to fourteen years and therefore longer-term solutions such as more electrification has often been ruled out. By providing in many cases rather grudgingly for the peak means a great underutilisation of assets at other times, but is inevitable.

The Peak and Natural Ways

As far as "The Way" is concerned natural ways are unlikely to get extensively overcrowded, there is plenty of sea and air for all the ships and aircrafts to move at any time, but port approaches, airport approaches and specialised routes such as the cross-channel ferries may approach saturation point at peak periods.

For the artificial ways such as motorways and railways, particularly near London Bridge which serves both Cannon Street and Charing Cross and Waterloo East the utilisation of the way is often very poor. Whilst this has often been commented about in terms of the railways many roads may be built almost solely to cater for the peak hours. The railways have to provide more track and terminals than would be necessary if loads more evenly spread. Cannon Street in South East London had eight platforms which has now been reduced to seven which are almost entirely used for the peak hours, and the station is usually closed on Sundays except when there are engineering works elsewhere which may mean that the station is then opened. Until recently it was also shut entirely on Saturdays. There are many similar examples both in the UK and elsewhere.

Poor Utilisation of Vehicles

Units of carriage and units of propulsion have to be provided in greater numbers in peak periods than if traffic could be more evenly spread. Operators can maintain many services on the Underground railways with a single power unit. In many cases, three or four carriages on the railways would suffice during the slack periods, but for peak periods operators have to increase the length of trains to eight or twelve carriages. There have been in the UK considerable discussions about the possibility of double-decker trains, which are used in other countries including the Netherlands. These have not been used recently in the UK, partly because the loading gauge is insufficient on many parts of the UK railways.

Poor Utilisation of Buses

Bus routes can provide a half hourly service in the non-peak but may need a five minute service when offices or factories are starting and closing for the day. Again, in some cases larger vehicles such as articulated buses have been used, where large numbers of people may have to stand.

The Advantages of Trams in the Peak

In the UK, as well there has been more recently an interest of the provision of tram services as has been shown in Croydon in London as well as Sheffield with the so-called Super Tram. The advantages of trams at the present time is that they nearly always have almost an entirely self-contained track so they have not competed for speed or space with other vehicles. Historically the tramways in Britain did not have this advantage, with minor exceptions such as some trams in Blackpool and also the Kingsway Tunnel in Holborn London.

Poor Utilisation of Terminals

Terminal facilities have to be large enough to cater for peak travelling periods, which result in under use in non-peak periods. A terminal is an interface between one method of transport and all the others so that the peak from one mode pours into another mode to create a peak there too. One can readily see this as almost any railway station or the ports or airports especially when a Jumbo Jet discharges its passengers into a frenzy of taxis, hire cars, private cars, the Heathrow Express as well as the Piccadilly line in London, only to be refilled by massive movements of people in the opposite direction.

Staff Utilisation

Another main problem is the poor utilisation of staff. Staff have to arrive before the daily peaks and have to finish after them so that staff hours are spread over longer periods than eight hours (the normal working day). This means that overtime has to be paid or there has to be a split shift system, which is unpopular.

There have often been problems of attracting staff because of unsociable hours. These difficulties were reduced when the UK had large volumes of unemployment reaching over three million in the nineteen eighties according to the published figures. This was probably an underestimate of unemployment so that staff were fairly easy to find, although whether they stayed in the job is another matter.

Lack of Systematic Regulation of Services

A particular aspect of staff utilisation is that transport industries are also very susceptible to accident problems and standards of awareness of safety regulations. This was seen in February 2007 when following the death of one passenger in Grayrigg, Cumbria, the railway system in that area was shut down for several days. This is in marked contrast to the road network where road accidents are far more numerous and kill far more people than on the railways. In turn, there is therefore a commitment to training people which requires staff to be of the right quality at all times. It is also noticeable that the railway drivers have to be trained before they are allowed to go over any new routes so that they know where the signals, points, etc. are. This is called route knowledge; the idea may be desirable but makes the peak problems more difficult since if there is a breakdown on one route which means that trains may have to be diverted it may not always be possible to use the same drivers on some diverted trains, since they do not have this route knowledge This is again in marked contrast to car drivers. On the other hand, both road passenger and road freight drivers receive more training than ordinary drivers.

High Staff Turnover

The problem of staff turnover arises partly from the problems of unsocial hours, which in turn raises both training and recruitment costs.

Solutions to the Problems of the Peak

Possible solutions can be subdivided into internal solutions and external solutions. With internal solutions, the operator solves the problems whereas with external solutions the problems are attacked by an outside authority and the solution is not within the control of operator. There have been many attempts to tackle the problems. This have included measures to diminish the peak, increasing traffic in off peak times as well as measures to reduce congestion at peak times by improving traffic flow, access to units of carriage, clearways and red routes on the roads as well as restrictions on parking and in some cases the provision of new Underground lines such as the extension of the Jubilee line in London as well as the extension of the Docklands Light Railway.

The East London line has been northward to Dalston Junction so providing a South East to North East London route without having to pass the central London area. Some of the Underground trains may be strengthened to have more carriages. Some of the other possibilities will now be considered. The Elizabeth line with a very frequent service is due to open in December 2018 and will ease some of the pressures on other lines.

Pricing Policy for the Peak on the Railways

An internal solution is to use pricing policy. The railways for a long while have charged different fares for the peak and off-peak periods. Until the 1962 Transport Act, there were statutory requirements about workmen's tickets at reduced rates, even though the costs to the railways were often large and the demand was inelastic, it being difficult to attract extra traffic that early in the day. On the other hand, presumably for commercial reasons there have been a few examples where using so-called early bird fares, the railways have tried to get people to travel on what is sometimes called the shoulders of the peak (i.e. just before the main peak times in the morning).

In many cases before privatisation the railways had no great control over their commercial policy, which was often dictated by political considerations. This is still true in some countries where in many cases season ticket holders have had far cheaper fares than would have been justified on purely commercial criteria.

Pricing Policy for the Peak on the Buses

Bus operators on the other hand have often had a system of standard charging per passenger kilometre at all times of the day. This has been criticised by many people including the late Professor John Hibbs in his book 'The Bus and Coach Industry' (1975). It remains to be seen how far greater technology such as the use of Oyster cards in much of London will be able to alter this practice.

Pricing Policy for Airlines

The airlines have had a fairly complex system of charging and to some extent the use of the internet can make this even more complex, since given the airlines want to maximise the yield on any one flight, there may be a range of charges depending whether the aircraft is getting near being fully booked or not.

Pricing Policy for the Peak on Road Freight

Road freight rates are often unpublished and many road hauliers have charged on a cost-plus basis, but there have been some examples of road hauliers charging more according to the peak, charging more in the peak periods and less at off peak times. Whilst it might be thought desirable for operators to charge full commercial costs, there may well be political pressures, which might try to prevent this.

Pricing Policy for the Peak and Political Interference

Politicians frequently will opt for what the economist would call the second-best solution, if economic constraints make the logical solution to a transport problem unattractive. There has usually been no rational system of charging for the roads in the peak, although since 2003 there has been a congestion charge, which has been levied in the London area on weekdays. This has been extended in February 2007. The system of road pricing was advocated in the Rueben Smeed report in 1964, but there has been reluctance by successive governments to introduce it. Rueben Smeed was a member of the then Road Research Laboratory. A poll online email petition in 2007 attracted one million five hundred thousand objectors to the idea of road pricing with some of the media supporting the protestors. It is not clear why road users should not be charged according to the value of the road. It is also noticeable that Norwich Union one of the giants in the motor insurance industry had originally introduced the concept of pay as you go for insurance policies (i.e. charging people according to the time of day, the type of roads that people will travel on). This is partly because accidents vary tremendously according to this. The data from such experiments would have been interesting not only for accident reasons but also to the people providing roads.

Company Cars

In the past, many car users in the peak have been travelling in company cars, and one estimate in the 1970s was that about 70% of new cars were company cars. There is therefore a considerable degree of perhaps unintentional subsidisation of peak time car users. Perhaps the main pricing problem is that pricing policies have not been applied in the same way, throughout any mode of transport and certainly not across different modes of transport.

Almost certainly commuter fares by rail would be higher on a commercial basis and similarly a road pricing system such as that used in Singapore would impose higher costs to the peak hour motorist. The EU aimed to charge the marginal social cost to the user. The concept behind this is likely to become even stronger as the EU in March 2007 had committed itself to reducing carbon emissions from vehicles including cars. The logic for this would be to prevent the second-best choices being made; the commuter by car would have to pay to compensate for the possible social disbenefits imposed upon them, such as noise, pollution, congestions and accidents. The problems of charging low peak fares led to court action over the so-called fair fares policy in Greater London in the early 1980s. The logic behind "Fair Fares" was that the charging of lower fares by both buses and Underground trains in the peak hours reduced social costs, including those of noise, congestion and accidents. Whilst the policy led to slightly higher local government rates (these rates have now been replaced by Council Tax), these might perhaps be justified on a cost benefit basis. The Law Lords understanding of the economic argument was not noticeable in their decision making.

Reducing Peak Travel

Action against off peak travel. The peak problem may in some ways be regarded as an off-peak problem in that if operators could obtain the same loads in the off peak periods the problems would not really arise. Operators may try to stimulate traffic in the off-peak hours. British Railways and its successors in the privatised railways have offered cheap day returns in the off peak, in order to try to stimulate demand. In the South East of England, they now offer tickets for three or four people for the same price as the two people travelling at the standard off peak rate. This, however, has not been well publicised. In a few cases in British Rail days they also tried to improve utilisation by offering excursion services at off peak times. Smaller bus operators have tried to improve off peak utilisation sometimes through excursion services or private hire, and in the 1980s on occasions London Transport used to run some of its double-decker buses, which would not have been used on Sundays to run trips to the seaside. Apart from fairly conventional methods of charging less during the off peak, the South East of England have issued a Gold Card to annual season ticket holders, which means that there is a discount for people at weekend times. Sometimes the railways have also issued Rover tickets for particular areas whether it is Devon, the South West, North West, etc. These have often been quite attractive but again have not been well publicised, and this has been even truer since privatisation since the revenue is gained by too many different companies.

The logic of Rover tickets, etc. which have also been used by bus companies often only on their own services is that if vehicles are not being filled, then the marginal cost of passengers are usually negligible since the cost of an extra seat is virtually zero apart from a marginal increase in petrol or diesel. The railways have issued a Family Railcard, which currently cost £70 for a three-year period and which gives discount on most off-peak journeys if at least one child travels. The requirement that at least one child must travel was to try to prevent commuters taking advantage of this for ordinary trips to work, when full revenue could be covered. Sometimes operators have used methods, which are similar used by manufacturers. In the 1980s, British Rail advertised extensively on Persil (a well-known detergent in the UK) to offer cheap fares for two people travelling together. The airlines particularly the low budget operators have been publishing their cheap fares, both in conventional newspapers as well as on the Internet.

Policies Aimed at Cost Reductions in the Peak Periods

Operators have tried to reduce the cost of running peak services. Bus operators have sometimes operated limited stop services. How desirable this is depending upon accessibility time for passengers as a proportion of overall journey time and to some extent to the type of passenger involved. Business travellers may be inclined to walk further compared with shoppers or travellers with small children, who may find walking to the bus stop more inconvenient. The advantages of limited stop services to operators is that they may be able to obtain more journeys from the same number of vehicles. The railways also use similar methods as when they run semi fast services. Trains outside of London often stop for the first time at Surbiton (about 12 miles, 20 km from the centre of London) to avoid holding up other trains on the very congested portion of the line between Waterloo and Surbiton. Road passenger operators have sometimes tried to minimise peak costs by having a service such as the Red Arrow service, which runs with a large amount of standing room, but relatively few seats. This means that not only are the buses are cheaper to construct than a typical bus, but unloading and loading is quicker during the peak. The use of automated tickets issues whether more recently at bus stops in the London area or has been the case for a long while with ticket machines on the Underground or the more recent introduction of the Oyster card has helped to reduce waiting time at stations. Another way of trying to reduce the cost of the peak is to try to ensure that the vehicle and in the case of railways, track maintenance is carried out largely in the off-peak hours. However, more recently this has often meant that rather than maintenance being done in the very early hours that the railways have often closed down at weekends. In the long run this may be undesirable, since passenger travel may be a matter of habit as much as anything else. Transport firms may use modern management techniques including operational research, queuing theory to ensure best utilisation of the vehicle. There would also seem to be considerable scope for doing this in road freight. Currently, there is evidence from Transport Trends that a large number of freight vehicles still travel empty on the return journey. If peaks do not coincide for all operators, it may be possible to hire vehicles in and out during the peaks. This is done fairly frequently in road haulage and to some extent can be done with road passenger transport where coaches may be used as buses during morning and evening peaks. The railways face a dilemma since whilst multiple units (carriages with a motive power unit in the end section) can give quicker turnaround in terminals in the peak than with locomotive services, the multiple units can only be used for passenger services and therefore the locomotive power is less well used than with locomotive hauled stock where the locomotives can be used for both freight and passengers. Peak efficiency may be improved by better utilisation of vehicles in the peak through speeding up the handling or loading processes. On the passenger side this can be improved by the increased use of travel cards, season tickets, Oyster cards as well as carnets (i.e. perhaps ten tickets being sold at a reduced price so that the boarding time on buses is reduced). Some bus services if not going as far as this, have had a no change policy, which helps to speed up boarding time.

Issuing Bus Tickets at Post Offices and Shops

In other countries noticeably Holland, bus tickets have often been issued at local Post Offices and this is now used in the UK.

Selling Carnets

In France, carnets of tickets which are transferable between users are issued at a much lower cost than if tickets are issued if at all on the bus.

Use of Unit Loads

Similar considerations have also arisen on the freight side where the provision of unit loads whether through palletisation, packaged timber or containerisation has helped to speed up loading and unloading considerably.

The Use of Articulated Vehicles

The use of articulated vehicles where the power unit can be used more intensively by leaving unloading and loading of the unit of the carriage to be done separately also helps to improve utilisation of both drivers and equipment.

The Split Shift System

Another possible solution to the problem of the peak is to have a split shift system, whereby people work two peaks, but not in an intervening period. This reduces peak problems but is usually unpopular with staff. Another possibility is the extension of the working day, to include overtime. This gives workers higher pay than the basic wage, but the system may not be universally acceptable and may lead to both disputes and absenteeism. The service may thus become more unreliable, since staff may be difficult to recruit in boom periods, when workers find it easier to obtain jobs elsewhere. The two-shift system is more expensive to the undertaking and workers who have to work alternate shifts may find the hours anti-social. A standard shift system ensures that all staff work either an early shift, a normal day or a late shift.

One rarely tried possibility would be to employ people solely for one peak. More people would have been available for this type of work, especially in a period of high unemployment, but almost certainly it would have been opposed by the trade unions. It has sometimes been used by rural operators in the UK and one operator used schoolteachers after school hours to operate this.

Casual Labour System

In the ports, the system of casual labour was used up to the late 1960s. Casual labour means that workers are only employed if work is available. This had many poor features for example instability of wages for the workers as well as little emphasis on training and employers are unlikely to obtain the quality of service they require, especially in a boom period where workers are less likely to be available. Currently many workers have complained about zero hours contracts.

Road Pricing and Alternative Schemes

Road pricing is one way of reducing problems by charging the motorist more for using the roads at peak times. Road pricing has the advantage compared with parking meters is that people who travel across a city would have to pay a contribution towards the congestion they cause. At present a through vehicle avoids parking meter charges even though it may cause double the congestion of a vehicle going into the town centre. It also has the advantage of greater flexibility potentially (i.e. that charges could vary according to the degree of congestion). The London system also has the advantage of charging less for vehicles emitting fewer fumes etc. Singapore has adopted the system and charges motorists entering the centre during the peak; it does this through a series of tickets issued from tolls. This ensures that the fixed cost of such a system are low (the cost of the booth), although the variable costs (the labour involved) may be fairly high. Since a system of road pricing is likely to be politically unpopular it may be sensible to ensure that alternatives for motorist are available for example Singapore has a park and ride system and also a bypass around the town centre so that travellers going from one side of Singapore to another can avoid the charges. At one stage Hong Kong was governed as a colony which had a different proposed system. Under this system, there were to be loops in the road, which would detect when motorists passed across certain roads in the peak hour. With this system, it would have been necessary to have a fairly sophisticated system so that when cars finally left the restricted area or where parked off road that charges ceased. The car or vehicle meter needs to be tamper proof. Both Hong Kong and Singapore road pricing systems were easier to administer since both countries were then small. Hong Kong is now part of China (formerly a colony of the British Empire). In Singapore, the majority of vehicles entering the centre are likely to be doing so on a regular basis and the number of overseas vehicles are minimal. This is because problems of providing suitable information to non-residents is more difficult. With the Singapore road pricing system, it is important to have booths on sufficient roads to avoid queues at the tollbooths, but not such a large number that administrative costs are high. Road pricing has the potential advantage of ensuring that people who currently use private car parks but cause congestion by entering the town centre will have to pay for what has usually been a free facility. This is in contrast to parking meters which provide additional revenue, as would road pricing but they have high administrative costs. The parking charges made do not necessary reflect the marginal social cost of the provision of roads. The original intention to provide off street car parking from meters proved in many cases to be undesirable, because it might encourage even more cars to enter the town centre. Another possibility is that of removing or restricting the number of private or public parking spaces. Since one of the assumed advantages of the car is that it can give a door to door service, such restrictions may well be effective in reducing congestion, although politically it would be controversial. Parking regulations, which often insisted on a minimum number of parking spaces for housing or offices in the UK were often unhelpful and these policies have tended to be reversed.

Poll Tax on Workers

An alternative approach is a poll tax on the number of people working in the city centres. The disadvantage of this is that it does not necessary reflect accurately the true social cost of congestion. It is undesirable since public transport passengers would still be charged.

Such a tax could also be regressive, since firms employing large number of low paid workers may well because little congestion compared with firms employing a number of highly paid workers, nearly all of whom travel to work by car. Also, firms which employ relatively little labour may attract large number of visitors who cause congestion. A further disadvantage is that unlike road pricing or parking charges, it gives no incentive to share vehicles, including cars which might partially help to solve the problem, or for the firm to provide subsidies to public transport users or bikes.

Car Sharing

Some commentators have suggested that car sharing should be part of the answer. Whilst most cars are designed for four or five people, in the rush hour in the UK and elsewhere, it is common to see cars with one or two people and comparatively rarely loaded to capacity. Some firms have however adopted a green travel policy, which tries to encourage people to use public transport, cycling or walking or if these are not possible to encourage car sharing. Sometimes car sharing is difficult because journey times do not coincide and there is also the absence of perfect information, although through the use of internet and other forms of communications this should become easier. Information may be improved on occasions during bus or rail strikes when local radio and newspapers have sometimes put people directly in contact with another making similar journeys. The cause of car sharing has not been helped by restrictions imposed by some insurance policies, whilst government policies which did not allow car owners to accept payments for such services has also not helped.

Improving Traffic Management

Improved utilisation of "The Way" could reduce costs. Roads could be designed so they have traffic flowing in one direction in the morning and in the other way in the evening. The Aston Express Way in Birmingham was originally designed with this concept in mind. The utilisation of the way could also be improved by encouraging the use of smaller vehicles. The "Cars for Cities Report, 1967" suggested that there might be segregation from other traffic partly from safety reasons and partly to improve flows of traffic for small cars. The House of Lords report suggested that small cars should be given parking preference especially for battery operated cars. The use of cycleways which might have generally higher capacity per metre width might also be helpful, and the Sustrans cycle network could be improved. Good road design can also help speed up peak movements. In the UK for example right-hand turns are often prohibited because they are cause of both congestion and road accidents. An associated advantage is a drop in pollution levels since traffic travelling at even speeds with even flows is helpful. However, the schemes may incur additional vehicle mileage and therefore increase total consumption of fuel, although higher speeds may help partially to overcome this problem. Traffic management schemes may restrict public as well private transport, although in some cases exemptions may be given to buses, taxis and cycles. An example of problems was that of Nottingham in the early 1970s where a scheme was introduced to give priority to public transport, but was later abandoned for political reasons. A good description of this can be found in Bendixson's book 'Instead of Cars' (1977).

Bus Roads

Schemes to give priority to public transport can improve utilisation of roads. The use of computerised traffic lights can be introduced partly for safety reasons and also to ensure faster flows of traffic. A comparable idea for the rail systems has been the introduction of coloured light signals rather than semaphore signals which has helped to improve both the capacity and speed of trains. There have been other proposals which have not yet implemented. One imaginative idea is that bus companies might be allowed to spend money on bus lanes or bus roads. The advantage of this is that the bus company itself could decide on priorities rather than relying on local authorities whose decisions are sometimes capricious. Bus roads and bus lanes generally give greater capacity measured in terms of numbers of people carried per metre width, since buses are much more economical of road space, especially as load factors in the peak are generally higher than for cars.

Fiscal Aspects

The effects of road pricing or better or cheaper public transport may have little effect upon people using company cars. It is difficult therefore to provide an optimum allegation of resources unless taxation of company cars reflects all benefits to the users and gradually there have been more restrictions on company cars in the UK.

Bans on Parking, Unloading and Loading

Whilst freight transport vehicles are usually in the minority in the peak hours, they will constitute a problem if either loading or unloading takes place in busy streets, and they may cause heavy congestion in towns, if their power to weight ratio is insufficient.

Bendixson's book 'Instead of Cars' (1977) suggested encouraging transhipment of loads so that fewer vehicles would serve city centres. Suggestions for municipal transhipment depots for firms themselves to handle transhipment have been made. Marks & Spencer have shown that the use of transhipment depots is environmentally helpful as well as being commercially successful.

Another possibility is to ban goods lorries at particular times of the day from loading or unloading. On many roads stopping is prohibited for all vehicles during the peak hours and a better flow of traffic is therefore possible. In the 1960s, a night loading scheme known as Operation Moonlight proved to be generally unsuccessful because of non-co-operation by some retailers.

Loading therefore took place over a longer period involving heavier costs. If such a scheme were to be tried again it would seem necessary, that all firms co-operated and if necessary this could be enforced by legislation. In the mid-1980s the Wood enquiry suggested that bans on lorries at weekends and during the evenings should be instituted, since these were the times when the noise of lorries most extruded in residential and other areas.

Improving Existing Facilities

One of the most common public answers to the problems of peak periods is improving the capacity of existing transport systems.

There have been suggestions that more roads should be built to cater for the increase in demand, although many people would query the idea that we should predict and provide. One problem is that the cost of building more urban roads is high because of the high opportunity cost of land. Therefore, it is likely that the law of diminishing marginal returns applies. Apart from high financial costs, new roads or road improvements are often unpopular with local residents. The absence generally of road pricing makes it difficult to judge whether or not new roads or road improvements are the most effective cost measure. In small towns where a large proportion of traffic is through traffic, bypasses may well be part of the solution. In larger towns roads such as the M25, which to some extent acts as a bypass are unlikely to have any significant effect on congestion, partly because of the large proportion of traffic to and from London which has its origin or destination there. Surveys would seem to indicate that the M25 is not used in effect as a bypass, but more to go from one part of London to another. This does not necessarily imply that the M25 is unhelpful, since it may give great benefits to users and give them access to congested areas of the most effective point of the system. There has in the past been a bias shown by local authorities to road building or road improvements rather than to public transport since road building attracted larger grants and often was paid for by central government. Before the introduction of Transport Policy and Programmes meant that until that time there were few grants to local authorities which could be used to help public transport. Railways could be better utilised if more private sidings were built. There has been more success however with private sidings in other countries, notably France and former West Germany. It is unfortunate that the Beeching Report in 1963 led to the removal of many private sidings from the centre of towns.

This may have been justified on commercial criteria but might well not have been justified on a cost benefit analysis criterion. It is possible to increase rail capacity through either longer trains or double-decker trains. Longer trains may cause problems in terms of signally and may cause addition expense in extended platforms. One possibility is to have more through trains as for example with Cross Rail in London now renamed the Elizabeth line, which would link Paddington and Liverpool Street. Double-decker trains would incur higher costs for bridges or tunnels. It may be possible to reutilise disused rail track. For example, Birmingham's Snow Hill has been reopened and Thameslink which passes through Blackfriars and Kings Cross has led to through services from North to South London. There have also been increasing use of light rail systems as for example with the Docklands Light Railway (DLR) from the Bank to Stratford and Woolwich which has been extended several times. Whilst the total cost of DLR which was estimated at seventy-seven million pound for sixteen miles of track and trains and a large number of stations may have seemed expensive, it would still have been cheap compared with extending the road system. Another possibility is the extension or building of Underground lines. Changes to the East London line have developed at the present time. The system has a high level of comfort. The high speeds have been obtained at the expense of accessibility to stations, the assumption being that most people would be able to get to the stations by car. The high comfort of course leads to great expense. In contrast, the light rapid transit system in Newcastle in the UK sometimes referred to as a tram system utilised what had been British Railway lines though with some new lines to the centre and its relatively low cost system which has since been extended. Both the Hong Kong and the Singapore rapid transport systems have developed new lines and the emphasis has been on capacity with Hong Kong trains carrying up to three thousand passengers. This is probably a sensible system in such countries where land space is obviously at a premium. The Croydon tram system was also one where existing railway lines have been used and to some extent the tram system in Birmingham also used some of the existing former rail track.

Cab Track

There have from time to time been suggestions for using new technology in the rapid transit field. For example, in the 1970s Hawker Siddeley a well-known aircraft manufacturer suggested the use of a cab track with a series of small cab type vehicles controlled by computer which would be running four to five metres above the ground and special track. It would have given great accessibility to city centres although there would have been some problems for passengers who would have to climb to the individual stations. One of the major problems of such a system would be the cost of having a completely separate track in built up areas rather than using existing road or rail track. In particular, there is a problem trying to find adequate space for such a system and some objections would have been raised on the ground that such a system would mean loss of privacy in areas where it operates.

Use of Flexitime and Staggered Hours

A more radical suggestion is that we try to minimise the peak by the use of flexi time. With flexi time people can choose for themselves when they want to arrive for work or want to leave for home. Sometimes the core time has to be worked so that management can rely on all staff being present for the core time, but would expect some staff to be out during their flexi time. A variation on this would be for some people to work from home perhaps for two or three days a week and only come into work again in effect around the core period rather than a core time.

Flexi time has many advantages for both travellers and operators, since the travellers can choose a system of transport, even if it is slightly unreliable since any lost time can be made up later. For the operators, it has the advantage of broadening the peak hours as some people will arrive early and some people will postpone their journeys until later. Staggered hours (i.e. arrangements where the times of work of different members of staff) will be fixed but not all start at the same time have been advocated, but unless there is a wide divergence between the peaks it is unlikely to have such a great impact upon travel patterns. Some large government departments such as the former Department of Social Security in Newcastle Upon Tyne arranged their times slightly earlier than usual to minimise the effects of the local transport system.

Changes in Land Use Including Changes to the Land Use Planning System

Other possibilities although long term ones include changes to land use planning. Demands for long journeys to work especially to capitals such as Central London could be altered if offices would decentralise. To some extent even the creation of the Docklands system has altered the nature of peak traffic. It becomes easier with the digital revolution since information can be available to almost anywhere and there is less need for large scale offices compared with the past, since information can be readily available almost anywhere. Some large insurance offices have gone further than this, by relocating some of their activities overseas. Successive governments have paid at least lip service to regeneration of inner cities and this has occurred in Liverpool and Manchester. This can lead to a reduction in congestion, although not if the jobs within the inner city merely attract people from suburban areas.

Subsidies to Public Transport

The widely-suggested policy at least by the public is that of subsidies leading to cheap fares so that more people travel by public transport. Such a policy has been tried in many Eastern European countries before the breakup of the Soviet communist empire in the late 1980s. Such a policy has also been tried notably with the then South Yorkshire Passenger Executive and to a lesser extent with the former Greater London Council. How far lower fares attract customers depends partly upon the price elasticity of demand. It depends as well to the extent which company car owners are subsidised. If the policy just attracts more people that would have otherwise just walked or cycled, this may be beneficial for other reasons but is unlikely to have any effect on congestion, although it may help safety by reducing cycling accidents. One longer term problem is that lower fares might lead to people making longer journeys to and from work, since prices of houses on the outskirts are often cheaper than housing in the centre, and potential commuters would no longer be deterred by high fares. Another problem is that if low fares merely attract off peak travellers this would improve utilisation of public transport, but would not necessarily lead to a reduction of congestion.

Conclusions About Problems of the Peak

We may conclude that there is no panacea for curing the problems of the peak. It is probably best tackled by a mixture of the methods outlined in this chapter, but it would no doubt continue to remain to some extent a problem in all forms of transport.

Above all a wide public transport understanding about the nature of the problem and the attempts of being made to solve will prevent the move ridiculous demands that are being made.

CHAPTER 23
Rural Transport

The rural transport problem has had considerably less attention paid to it than the urban transport problem. Indeed, some people would question whether there is in fact a rural transport problem in developed countries such as the UK.

High Volumes of Car Ownership

In some areas such as Central Wales over 90% of households have cars. The question therefore might be asked is why cannot cars meet all passenger transport demands? Part of the answer is the number of people who do not have access to cars, this includes for example in the UK the under 17s who by definition are not allowed to drive them, the elderly who face the choice between no public transport or a very sparse public transport system or having to move away from the area in which they were brought up. It also includes those who are sufficiently disabled not to drive, as well as a small number of people who for whatever reasons do not wish to drive in the first place. However, the number of older people with driving licences is increasing and there are also increasingly more women with driving licences so that the problems for the elderly widow is likely to decrease slightly in the future. On the other hand, since people live longer this may only be a slight effect.

Problems for the Physically Handicapped in Rural Areas

There are also problems for the physically handicapped. There is perhaps the more hidden problem that even where there is a car in the household that this does not mean that all members of the family have access to it. Another hidden problem may be that some people particularly those in less well-paid jobs may find that given the rundown of the public transport system in rural areas they have to join the urban drift, since they cannot find jobs which pay enough money to cover ordinary household expenses as well as the cost of running cars.

Original Widespread Railways in Rural Areas

Historically the railways served most rural areas so that in the nineteenth century only two villages with a population of over 3,000 were not within three miles of a railhead. However, the provision of rural railways is very expensive, since the fixed costs are high and with limited demand average costs will also tend to be high.

The railways could generally only provide an infrequent service. Also because of gradient limitations the railways frequently did not serve town or village centres. Centres were not served because of the terrain, in some cases town or village centres would have meant either expensive diversions, tunnelling or the destruction of buildings. In rural areas, as with the urban areas transport costs were often unnecessarily high, because they had to pay large sums of compensation to landowners.

Community Partnerships

Some of the lightly used railways would originally be designated as community partnerships and if this was not successful then they would be closed. There have been some community projects such as Swale Rail which covers the lightly used Sittingbourne to Sheerness line in Kent and also the Medway valley line.

Because of the infrequent rail service and because accessibility was often poor in the 1920s and 1930s bus services often offered a more convenient though somewhat slower service.

Reductions in Staff Costs Whilst Still Giving a Service

The stations in many areas have minimal staff in an attempt to reduce costs or rural services. Often there are now unmanned stations. This reduces the quality of information to the passengers though sometimes this has been overcome imaginatively, for example through the use of remote controlled loudspeaker systems which announced train arrivals on the Hereford-Shrewsbury route in the 1980s as electronic equipment has become more sophisticated it is possible to give details via computer generated screens to show when trains are arriving and even how late they are, etc. even at unmanned stations. There can also be help buttons in unmanned stations so that people can check whether the trains are running and how late it will be if there is not money for a loudspeaker system. Messages can be sent directly to mobile phones. This does give some reassurance to passengers. Apart from the costs of providing the stations, stopping and starting costs more in fuel and also reduces overall speeds. One of the features of some rural railway lines is therefore the provision of halt signs that the train will only stop by request to the guard from passengers on the train or by a request from the station itself. An alternative approach to unmanned stations would be for station staff to issue more than just tickets for example selling other items. Alternatively, it might be possible to have shops which issue rail tickets at the station. This would be similar to the provision in many countries of small shops issuing carnets (i.e. books of tickets for journeys). This was tried to some extent by the former rail operator for much of Kent, by Connex South East in more urban areas before it lost its franchise.

A more conventional approach is to try to reduce the costs of the rural services. British Railways introduced conductor guards who collect and issue tickets on route. If guards are to be used anyway, this gives rise to little extra cost and has the additional advantage that passengers often prefer the more personal service and can ask questions. The fears of using public transport seem often to have been overlooked. The more radical step of having the driver issue tickets as happens on buses has not, however, occurred in this country. This has occurred on some rural routes in Austria. In the UK, there would almost certainly be strong trade union objections to this and there could be risks to safety.

The railways have used centralised traffic control as well as radio control to try to reduce the costs of signalling. The use of a signalman has also fallen as far more unmanned level crossings have been used or in some cases level crossings eliminated through the use of bridges. Large numbers of rural stations have been shut.

Rural Railways

Part of the problem with the rural railways was that the revenue of the rural stations was looked at but not the revenue from tickets to such destinations. Therefore, rural ticket sales shown for a small seaside resort might well be small, but this would not prove that the demand was low since many passengers might book their tickets from other areas to the seaside resort and return.

The use of such lines as feeder services was often overlooked, the assumption being that people would merely get cars to the nearest main line service. In practice, in many cases people did not do this and so the revenue lost was greater than that which was expected. The Serpell Report 1983 suggested a wide variety of options, under options A and B which would have reduced the passenger mileage opened to 7,900 miles and 9,800 miles respectively all branch lines would have been closed and therefore the only rural stations which would have remained open would have been those where the main line ran through rural areas.

Using Voluntary Railways

Another approach would be to allow the greater subsidies to the voluntary railway societies which currently run a number of services in the UK usually steam operated since these arouse the greatest enthusiasm. The voluntary societies have the great cost advantage of not generally paying any salaries though time they do pay low salaries to certain members of staff. Whilst there are relatively few examples of voluntary railway societies running services throughout the year they may be willing to operate services at weekends when demand arises for leisure services and also at holiday times when there is a likely influx of tourism. The Swanage line service has operated as a park and ride services to avoid congestion in a very busy area in the summer. In West Yorkshire, Northern Rail would like to use the track maintained by the Keighley and Worth Valley steam railway so that Northern Rail could provide peak hours' services to reduce road congestion in the area.

The high fixed costs of rural railways have meant that in most countries including Britain there are relatively few pure rural routes though some rural areas are served by services between two major towns. One of the problems which many rural services had is that demand is very seasonal for example, holiday makers come in the summer, but there is very little traffic in the winter months. Unless casual labour such as students are used to cover these seasons, labour has to be paid for all the year.

The use of voluntary railways may partially help to overcome this problem. Ironically, in spite of the fact that British Rail found it difficult to attract staff for steam engines, voluntary railway societies seem to find it easier to attract people who will work on steam trains in their spare time. This seems to be almost a British phenomenon. It may also reflect a local interest in relatively small organisations compared with a large organisation. The novelty aspect usually enables the voluntary railways to charge a much higher fare than that which British Railways or the privatised railways would have been able to obtain. It is noted more however in recent years, that the voluntary railways too have used some form of price discrimination in order to gain the maximum revenue. There are usually additional receipts from sales of other items including books, gifts, etc. However, at the present time most of the voluntary railways serve relatively little transport purpose partly because of the lack of through running between the National Rail network and the voluntary railways.

The Wensleydale Railway

An unusual approach to the rural railway has been that of Wensleydale. This runs currently from just outside Northallerton on the East Coast main line from Leeming Bar to Redmire. There is a bus service from the rail network to Leeming Bar and it has ambitions to be able to connect in the main rail network at both Northallerton on the East Coast mainline and possibly Garsdale on the famous Settle Carlisle line. Unlike most of the preserved railways, it is not relying generally on steam trains but uses modern diesels. These are low level trains so that there is no awkward step into the trains which means that it caters better for disabled people than many preserved trains or even some conventional services. It has raised money through shares, but the incentive is free trips on the railway rather than conventional dividends. It uses existing railway track which for a long while was solely for military use. It does as with most of the preserved railways depend partly on voluntary labour. It hopes eventually to be able to go through to link up with the Settle Carlisle railway which could prove to be a useful diversionary route on occasions.

It also hopes to be able to serve tourist towns such as Hawes which at the moment has virtually no public transport, although it is a major tourist centre as well as being known for dairy products, etc. As with the preserved railways, it tries to use the stations as being part of a selling point for the usual memorabilia, etc.

Reducing Costs

The use of Pacers and Sprinters introduced by British Rail in the late 1980s to replace the ageing diesel multiple units and most of the old diesel trains, should have had the advantage of higher speeds and lower maintenance costs thus increasing both revenue whilst decreasing variable costs. The rural services in some cases later became a separate organisation within British Rail, and this had the advantage that they were no longer reliant on stock being handed on to the rural areas which was also called cascading.

The Railways in Rural Areas

The railways in rural areas were subsidised under the public service obligation which was introduced under the 1974 Railways Act. The public service obligation did not only cover rural railways. Prior to 1974 rural rail services had been drastically reduced following the publication of the Beeching Report in 1963. Whilst some of the services almost certainly were unprofitable by any standards, mistakes seem to have been made. The revenue from individual stations transport costs were often allocated on the average cost bearing in mind the number of services. Therefore, if a station had 40 trains per day serving it and the feeder line had eight services per day, 1/5 of the costs were allocated to the branch line. However, often if the branch line were to be closed all the costs of keeping the station open probably remained unchanged.

Growth of Car Use and the Decline of Public Transport

In the post war years, the car offered a more convenient form of transport than either bus or rail services and the percentage of car ownership in the UK is much greater in rural areas than in urban areas with similar incomes. The low marginal costs and even lower marginal perceived costs of motoring meant that this was usually much lower than that for bus or rail services so that the vicious circle of decline in public transport causing higher fares tended to go on.

Whilst the provision of public transport in rural areas has been a problem other problems such as the high cost of providing roads in rural areas should not be overlooked. There is almost certainly as well some degree of cross subsidisation for the postal services with the urban areas subsidising the rural areas. If the post office were allowed commercial freedom, it would almost certainly wish to cut back on rural postal services or alternatively raise postal charges. This is part of the problems which the Post Office faced in 2007 as it lost some of its monopoly power. It also faces similar problems now that it has been privatised. Almost certainly the private operators will not wish to run rural postal services.

Use of Subsidies

One possible solution is the use of subsidies. Following local government re-organisation in 1974 many counties for example Cumbria in North West England spent increasingly large amounts of money upon rural transport subsidies. Other counties, however, spent considerably less sums on public transport, the Independent Transport Commission in their book "Changing Directions" (1974) published by Coronet Press complained about bias in certain counties, for example so that much more money was spent on provision on urban parking in towns than on rural public transport needs. There are difficulties with the use of subsidies especially as the present government has cut local government expenditure. There are also the practical problems of trying to decide what types of public transport provision if any should be subsidised, for example should the subsidies be only going to people travelling to and from school or business or should the subsidies extend to shopping trips, etc.

Subsidies for School Usage

The closure of many smaller schools in rural areas has meant greater distances travelled by students. This in turn means the local authorities having to provide services, the costs of which are likely to be borne by the Local Authority. Subsidies in any case are frequently paid to pupils in rural areas since there is free transport for journeys over three miles for secondary school children. This is much more likely to apply too rural rather than urban areas. The Conservative government in the early 1980s tried to eliminate this subsidy by charging fares but public pressure proved too great. Similar problems arise with the provision of health care, the closure of many so-called cottage hospitals in the 1970s meant that distances travelled to and from hospitals were much greater for many patients and perhaps as importantly for visitors to such patients.

Under these circumstances even car ownership does not necessarily provide an answer if the spouse is either unwell or cannot drive the car. A cost benefit analysis approach as suggested in Chapter 17 might have shown that sometimes the reduction in costs to the NHS were lower than that of increased costs to the consumer.

There has been little logic in some cases, for example the magazine Modern Railways has complained that North Yorkshire county council has switched children from trains to buses on safety grounds even though rail is obviously safer than buses on the Whitby line.

Subsidising Cars

A more radical suggestion which has been made from time to time is that it might be cheaper to provide cars for the relatively small minority of people in rural areas who do not already have them. This has sometimes been rather glibly termed 'The Volks' for the folk's approach. In extreme cases, former British Rail calculated that it would be cheaper to do this rather than to continue the service for existing passengers. The underlying problems, however, of young people travelling to and from school and the elderly would still largely remain. An alternative approach is giving subsidies to car owners who would be willing to undertake either ad-hoc or regular journeys for people who might need it, for example for shopping or hospital purposes might, however, eliminate some of these problems. There is, however, here the psychological point that people who own cars do not necessarily wish to share them and equally the fear of many people of accepting lifts from other people. However, in some areas there are voluntary car services for people to and from general practitioners and hospitals, etc. which can be very helpful.

Solutions to the Rural Transport Problem

An alternative to subsidising public transport would have been to have subsidised some of the rural activities. This has already been done to some extent for example, some of the post offices in rural areas will not pay their way and are therefore currently under threat, the smaller schools may be more expensive to provide than larger schools. Whilst the subsidising of activities may seem odd it should be remembered that the purpose of transport is to find access to facilities and transport has little if any inherent virtue in itself. When considering school closures, it would be sensible not only to consider the costs of education per pupil as well as the quality of education but also the higher cost of providing transport.

More Imaginative Use of Buses

It has been suggested that the bus operators could be more imaginative in their form of service, for example running to different towns each day so that they at least give a once a week service for shopping and other purposes. The concept of dial-a-ride has also been suggested. This was used mainly in towns in the 1970s. It might, however, be suitable in the rural areas where some villages are often off the main roads and to serve them all the time means a slow cumbersome and costly service. If, however, they only go to these villages when there is a specific demand, it would speed up the service and also provide a more attractive cheaper service to other passengers. The problems, however, are that those people who are without cars will sometimes also be non-phone owners, although this has become much less true in recent years as the cost of phone services has been reduced. An adaptation of the service might, however, therefore be that people would put in their demands for the service, for example to the sub post office the day before the service was due to run. Although it may seem strange little attention seems to have been paid to potential passenger flows in some of the rural areas. In the mid-1970s for example Hertfordshire County Council carried out surveys in particular parishes to try to identify particular flows.

One problem, however, is that people tend to overestimate the potential demand and some transport co-ordinators within the rural counties have adopted a rule of thumb (i.e. that public transport operators can anticipate about 50% of what people say is the likely demand). Changing patterns of shopping and population mean, however, that it is important for transport operators to keep up to date with likely demand.

Reducing Costs of Buses

Reducing costs has not been confined to the railways. One-person operated buses have been used everywhere in rural areas as drivers and conductor's wages have in the past been a very large percentage of total costs. The use of one-person operated buses has perhaps been more beneficial in the rural areas, as proportionately less time is taken in picking up passengers than in urban areas. There have also been moves away from the conventional larger double-decker and single-decker services.

There have been suggestions that mini buses could be used much more frequently, these have a much lower initial cost than conventional buses and it is easier to find and recruit bus drivers. The breakup of the National Bus Company (NBC) means that in rural areas there will have been downward pressure on wages to reduce costs.

Use of Minibuses

Mini buses have not only been used by conventional bus companies but have also been used very frequently by community organisations. There could be significant advantages, if mini buses could be horizontally integrated for example if organisations which otherwise need small vehicles would run them in the peak with perhaps drivers only being used for that time. This might be possible, for example with garages selling petrol and using mini buses both for passengers and also to carry around small items. One of the problems at the present time is that there is really no specially designed dual purpose mini bus which could carry both passengers and freight and which will lead to better vehicle utilisation.

Passenger Travel Combined with Freight

However, an example of passenger travel combined with freight is the use of post buses. This has been extensively used in other countries including Switzerland partly because of the lack of provision of alternative transport services in such areas. Within the UK it has been used considerably in rural parts of Scotland and also some areas in England including the Isle of Wight. The introduction has been limited partly because the Post Office has not run such services; if there has been any other form of public passenger transport, however, inadequate. It has been claimed that the post buses have sometimes been profitable though artificially because post buses have received in the past grants as passenger service vehicles. This grant, however, was phased out during the early 1980s. One of the advantages of the post bus service was that the Post Office found that if the vehicle had passengers that the driver was likely to drive more carefully so that maintenance costs were reduced.

Cross Subsidisation

Up until the late 1970s many of the bus services in rural areas were cross subsidised by subsidiaries of the NBC and municipal companies from the urban services. However, the combination of the 1980 Transport Act and 1985 Transport Act put an end to this practice.

Smaller Companies as the Answer

It has sometimes been suggested that large organisations such as the former NBC has not been the right one for rural areas. This is partly because of the distance away from headquarters of most of the services which makes it more difficult for the management, however, good to be able to gauge the local needs. Another point is that the NBC had to pay national wages which are perhaps higher than those which a smaller local company would have had to pay. Small local firms may sometimes be able to run the bus service as an adjunct to other services, for example in conjunction with a local coach service or contract hire work. In such small firms, there may well be more flexibility of work for example the bus crew may do other types of jobs during non-running periods. One possibility may be whether it be a former National Bus Subsidiary or a small private company may be to have a more flexible workforce. One of the problems in rural areas is high fares because of high average costs. Bus operators will not usually wish to come into the business unless they are sure of potential demand.

Community Buses

Community Buses have been used in some rural areas such as Norfolk. The logic of community buses is that crew costs are a fairly high proportion of total costs. They may be slightly higher in the UK at the present time, since there is usually a national wage structure which may mean that wages would be higher than that otherwise payable in the rural areas. Even with this higher wage however it may be difficult to attract bus drivers who are not car drivers because of the problems of being able to get to the local garage. The use of community buses whereby voluntary drivers will undertake the work at first sight therefore seems a logical answer to part of this problem. However, a number of problems arise in practice. The first is that in such areas even though crew costs are a large part of total costs the revenue may still be insufficient to cover the costs of depreciation, maintenance, etc.

The second is that whilst it may be possible in the short run to find voluntary drivers (e.g. in the summer, it may be more difficult to find a pool of people who could be relied upon to do the job throughout the year). There is also the problem of arranging suitable training facilities. For this reason, community buses may be helped if the local bus operator, realising that there is little scope for the provision of such services gives assistance as the former NBC had done in some areas.

Bus Clubs

One of the problems of rural transport is that fixed costs whilst not high in relative terms compared with the cost of rail are sufficiently high that even to merely break even a high fare may be charged if demand seems likely to be insufficient. If, however, demand could be stimulated in some way then lower fares could be charged. The operators may, however, be unwilling to experiment in this direction. For this reason, in some rural areas there have been attempts at bus clubs where people from a particular area or village or over a potential bus route guarantee to become members in advance. This has the advantage of guaranteeing the operator money whilst at the same time giving a relatively low average cost and zero marginal cost to the passenger. If sufficient people are interested in forming a bus club perhaps initiated by the local parish council, then the average cost for the relevant season ticket may be much lower than if they were to be charged for individual journeys to be successful it needs either a voluntary body or perhaps a parish council to get together to find potential customers. It is unlikely that unless the bus operator is very local that they will be aware of the potential passengers. Usually with a bus-club additional passengers can travel as well though the fare for these will usually be relatively high in order to encourage more people to join the bus club.

The operator being guaranteed this fixed sum in advance will find it easier to make suitable arrangements for both scheduling and for drivers rather than facing the usual problem of uncertain demand.

Greater Publicity

One way of trying to increase demand rather than reducing costs as such is to publicise the routes more efficiently. Some rural county transport co-ordinators have been very good at this. The mere production of an overall timetable may well be useful to passengers especially if more than one operator is involved. This may be particularly useful to tourists from outside the district.

Almost paradoxically the national British Rail timetable sometimes gave some details of bus operators where railway lines have been discontinued. This assisted British Rail, since it gave greater information to potential rail users. Unfortunately, the national timetable which had continued under privatisation was discontinued at the end of 2007, although there has been a new edition published by The Stationery Office.

Swiss and Liechtenstein Example

In Liechtenstein, it is possible to buy zonal tickets from a neighbouring country (i.e. Switzerland and then the main bus route from the station and then the feeder service connects with another with the bus driver of the bigger bus looking out for the smaller bus).

Voluntary Drivers and Cutting Costs

Another method of reducing costs in rural areas would be the use of voluntary drivers for local services. This has been tried to some extent in East Anglia. The problem here is cost of training of the voluntary bus drivers and also their reliability. People who may volunteer at a meeting on a summer's evening may be more reluctant to turn out on a cold winter's morning when the need for the service may be much greater.

Use of Rover Tickets

The use of Rover tickets covering a wide range of bus and in some cases rail services as for example in North Wales may be helpful. To the passenger this gives a zero marginal cost once they have bought the ticket and to the operator it may increase revenue since in most cases the services are underutilised. It may also encourage tourists to come to the particular area in the first place. Rover tickets have never been well publicised and the fragmentation of the rail network has made it even more difficult to find details of such tickets. The use of more imaginative fares policies may be helpful. Since the Cambrian Coastline in Wales was under threat in the early part of the 1970s, there have been vigorous attempts to increase patronage partly through the use of Rover tickets but also through an increasing number of excursions, evening fares and so on. A more recent example of this has been the Leeds Settle Carlisle Railway which had been under threat partly because of the sparse population which it serves but also because the cost of maintaining some of the viaducts. Since that time when it was announced that it was to stay open, there have been strong attempts by the local community to look after some of the stations and a few stations have been put back into service. In 2007, Dent Station announced that it could be used as holiday accommodation as a base for the area even though the village of Dent itself is four miles away.

Cost of Road Provision

Concern has been expressed from time to time about the sheer cost of road provision. There have sometimes been suggestions that County Councils should abandon the maintenance of some of the less utilised roads in an effort to avoid costs. This is unlikely to endear itself to local residents but would be in line with some of the more extreme commercial thinking.

Use of Taxis

For some people, particularly visitors to the area, taxis may be part of the answer. Currently taxis have been very expensive on a mileage basis though if they are used infrequently and the demand for total journeys is small they still might be cheaper than owning a car. They may well be used by visitors fairly frequently for example within the Lake District in North West England but are unlikely to be a solution for regular travellers especially for long distances.

Subsidies to the Ferry Companies

Not all the problems in rural areas are necessarily solved by rail or road services. The Islands traffic may require subsidies to the ferry companies notably MacBraynes in several parts of Scotland. The use of roll-on/roll-off ferries has helped to improve the utilisation of such services and in many cases additional revenue can be obtained from operating mini cruises (e.g. at Easter time when ordinary commercial traffic is unlikely to wish to travel).

Provision of Air Fields

For longer distance journeys the provision of air fields may be a relatively cheap form of transport. This is especially likely to be the case where journeys would otherwise involve long distances to avoid rivers, marshes and so on. The cost of some of the smaller planes is not necessarily very substantial.

Greater Publicity by Rail Services in Rural Areas

Sometimes where there are sparsely used services the railways could give more details of the attractions, particularly walking perhaps from one station to another or in a few cases even going from one route to another which would have the advantages compared with the car, of making a linear rather than a circular walk possible. In some cases, as on the Settle Carlisle line, there have sometimes been guided walks from the stations, different lengths of 6–20 miles. The advantages of this to potential passengers is that often people may wish to go for walks but find trying to use even Ordinance Survey Maps which are otherwise excellent, quite difficult to try to estimate the length of time required. They may also have difficulties in trying to follow routes from the typical guide books, especially where features may change even over a relatively small time. Often the guides themselves have good knowledge of the area and can give information, which would not be available to the conventional passer-by.

Railways Sending Sales Representatives to Schools

One possibility, which could be revived, would be for sales representatives to go to schools, as happened in the 1960s when in many cases large numbers of children were attracted to rail trips. The advantages of this from a schools' point of view is that it usually avoids the sickness which is unfortunately only too common on coach trips and also it is easier with modern railway carriages to be able to see more easily what is happening plus the advantages of toilet facilities. Some of the community partnerships have revived this practice.

The Welsh Highland line, which runs from Caernarfon to Porthmadog in Snowdonia the steam train line which is the longest regular steam service in the UK. Whilst it starts and finishes at sea level, it climbs to a maximum height of 650 ft. On much of the route, it uses old trackbed most of which was abandoned in the 1930s which reduces the cost and again the novelty feature will mean that it can be used from there to go on to the Ffestionog line which is a major attraction in its own right. It also provides a service for people in an otherwise poorly served area for public transport. In December 2017, Julie Walters the well-known actress travelled on the line during a programme about coastal railways which would have given more publicity to the service.

Hadrian Wall Service

Sometimes there is also the possibility as with the Hadrian Wall 122 bus service referring to the date when Hadrian's Wall was being built, which provides a guided commentary to an area which is otherwise inaccessible and people might prefer to be able to travel in a group with the social interaction involved.

CHAPTER 24

The Demand for International Transport

The demand for international transport is similar to domestic transport. Demand for international transport will depend upon the relative prices of producers and upon consumers' incomes and expected incomes. The relative prices, however, will be affected by changes in exchange rates, by tariffs (customs charges) and different taxation systems, for example value added tax in different countries. The relative prices would also be affected by the relative rates of inflation in particular under a system of fixed exchange rates such as that which existed up to the early 1970s. Fixed exchange rates meant that the pound was fixed within narrow limits to other currencies such as the USA dollar and most Western European nations. People could therefore know when they were going to other countries within very narrow limits what prices were going to be. The same applied to buying goods from abroad. There were exceptions when for example the UK devalued the pound sterling in 1949 and 1967.

In an era of floating exchange rates which has existed since that time for the UK (although not for the countries which now have the Euro as their currency), for example in the 1980s, the pound sterling fluctuated between £1 equals $1.1 and £1 equals approximately $2.40. This would have meant that a car costing £5,000 in the UK would have cost (ignoring transport costs) $5,500 in the USA, at one period but would have cost $12,000 at another period. The effects of such fluctuations are not confined to the demand for freight. The rise in the price of the pound will tend to deter overseas tourists visiting the UK whilst low values such as that of £1 equals $1.1 would have meant that more tourists would have wished to enter Britain. Such fluctuations in the exchange rates can pose problems particularly for freight operators, since they can find problems of backloading become more acute. For example, if the price of the pound is high, there will be relatively few goods exported, but there will be a considerable demand for the transport of goods entering the UK. The Euro which is now used in Austria, Belgium, Cyprus, Estonia, Finland, France, Germany, Greece, Ireland, Italy, Latvia, Lithuania, Luxembourg, Malta, the Netherlands, Portugal, Slovakia, Slovenia, within the EU removes one more obstacle to international trade especially in the continental countries where for most trading purposes, there is little difference between national and international traffic. Within the Eurozone countries which has existed since 2002, there are no different exchange rates for countries belonging to it and so trade is unhindered by either transaction costs or by the fluctuations of the currency. Transport economists observing the differences between the rates at which people can buy the Euro and sell it in terms of pound sterling, will see that the difference may stop some people going to and from the UK to France compared with for example people going from France to Italy where there are no transaction costs.

The Effect of Tariffs on Demand for International Transport

The demand for international transport also depend upon tariffs. Within the EU, the absence of tariffs on manufactured goods will lead to an increase in demand for both imports and exports from EU countries to the UK. This is known as trade creation. The common external tariff (CET) which means that goods from non-EU countries pay the same customs duty whether they enter Germany, Denmark, France or any of the other EU countries will mean that since 1973 non-EU countries such as the USA, prices will be slightly higher compared to EU prices. British exit from the EU will alter this.

Tariffs cause trade diversion and may be important for ship-owners trading between France and the USA. How much the typical CET of 5% to 7.5% will alter international trade will depend upon the price elasticity of the goods concerned. Economists might expect the demand for raw materials to be more inelastic than the demand for manufactured goods where there are often a much greater variety of competing goods.

In the UK, value added tax is not imposed upon exports and therefore the price of UK goods has been lower for an non-EU tourist than for a UK buyer. This applies to the other EU countries. There has been considerable press and public controversy about the harmonisation of value added taxation across EU since the creation of the single market in 1992. Currently, there has been considerable debate about whether Britain can remain in the single market following Brexit. The reasons for the harmonisation would be that otherwise there might be a distortion of international trade.

The demand for international transport will depend upon relative incomes in other countries. In the 1960s, there was very fast growth in most of the then common market countries and the demand therefore from these countries for goods and services rose considerably. Currently the rapid rise and size of the Chinese market has meant that there are more opportunities to both receive imports from that country as well as to export goods and services. The Beijing Olympics and Paralympics 2008 added further opportunities.

Similarly, the London Olympics and Paralympics 2012 added more opportunities, as many Chinese people would have both watched the sports in the UK as well as television and would have been influenced by the spectacle. The Brazil Olympics and Paralympics 2016 had the widest media coverage ever and many people would have travelled to that country affecting both tourism and the transport industry generally.

Normal and Inferior Goods

Economists classify those goods and services for which demand is lower as income rises as inferior whilst classifying those goods, which rise as income rises as normal. Most manufactured goods are normal goods and therefore as incomes rise throughout the EU, economists would expect the demand for manufactured goods to grow and therefore the demand for transport of manufactured goods between the member states to grow accordingly. The credit crunch 2007 onwards has reduced economic growth.

However, the demand for food is unlikely to rise (at least in volume terms) if incomes rise. The demand for transport of food will be affected by the Common Agricultural Policy (CAP). The CAP aims at self-sufficiency of the EU in temperate climate products. This has therefore affected the demand for example for New Zealand dairy products and will have led to a decline in the transport of those products. The demand for sugar from the West Indies will also have declined as this competes directly with the demand for EU sugar beet, which has been heavily subsidised. If the countries, which have signed up to, the climate change convention take it seriously, then agricultural policies will change in the EU, the UK and the USA.

The Quality of the International Transport System

As with domestic transport, the quality of the transport system will affect demand considerably. The demand for international transport rose considerably in the nineteenth century with the development of refrigerated ships, which meant that foods were available from a much wider range of destinations. More recently, the development of faster aircraft has meant that there have been a much wider range of goods including fruit and vegetables for European consumers. The quality of international transport has also affected the international business travel market for example the introduction of Concorde on a variety of routes meant that it was much more possible to make visits to other countries without using up many working days.

The advent of the Dreamliner and the European airbus will also reduce the time taken. However, the demand for such travel will be affected considerably by the availability of substitutes particularly in the realms of telecommunications such as email systems and computerised systems. Email will reduce the volume of international post, as it gives international communications cheaply. Perhaps the most conspicuous effect of improved international transport has been the rapid growth of tourism. For more details about this, interested readers can read the book written by David Spurling and John Spurling entitled "Tourism Matters: Study Guide for BTEC Travel and Tourism" (2018). The volume of tourism has risen because of increases in incomes, reductions in prices and increased leisure time. The sheer speed of modern aircraft has meant that a much wider variety of destinations has come within ordinary people's availability. This led in the 1950s and the 1960s to the rapid increase in demand for tourism to Spain, Portugal and Italy. Currently many people travel further as aircraft have become faster. Therefore, travel from the UK to Australia and New Zealand has become common particularly in January and February when the climate will prove more of an attraction and similarly the longer days in July and August will mean more people willing to travel from the Antipodes to the UK.

Shocks to the System and Effects

Shocks to the system such as the World Trade Tower tragedy in 2001 or the tsunami disaster in Boxing Day 2004 may alter the pattern of tourism as people temporarily avoid some areas. Terrorist attacks in Europe, Syria and other Middle East countries will affect demand. Relaxation of tensions between the USA and Cuba will have increased travel and trade between these countries. However, it remains to be seen what the current President of the United States Donald Trump would do about this.

Standardisation of Size of Consignments

The increase in international transport has both been the cause of and the result of improvements in international handling systems. International transport has been facilitated by the use of the ISO containers which have been of 8′ × 8′ by a multiple of 10′. More recently, there have been wider ISO containers which have a width of 9 ft. 6 inches. The standardised container size has meant that handling speeds have become increasingly faster. Specialised cellular container ships have reduced significantly the time that ships take in ports and hence the ship-owners' costs.

It has also reduced the risk of damage and pilfering whilst handling the goods and so therefore indirectly the price to the consumer has fallen as insurance premiums have fallen considerably.

Other Handling Features and Changes Affecting Demand

The use of roll-on/roll-off ferries has also increased considerably for example between the East Coast Ports of the UK and Continental Europe. Ports such as Felixstowe which are reported in the 1960s to have been sold for a few thousand pounds have developed rapidly. Dover has been the premier port for roll-on/roll-off partly because it involves the shortest sea route between the UK and the continent. Some doubts however have been expressed about the stability of roll-on/roll-off ships especially since the Zeebrugge disaster in 1987. The roll-on/roll-off ports have also gained in the UK from domestic transport improvements for example the construction of the M25 (the circular road around London) has considerably reduced the time taken to go from the Midlands to and from Dover and Folkestone. In the early 1990s the pessimistic views of the Sealink owner about Dover being a dead port if the channel tunnel was built have not actually occurred and instead Dover has taken on a new role as a cruise port and it currently claims to be the largest tourist port in the world.

Location of Industry

The demand for international trade will be affected by the location of industry within particular countries. In the European continental countries such as France often the distances between regions of particular states can be much greater than many international movements. For example, transport costs will generally be lower between parts of Germany, Holland or Belgium than between Southern and Northern Germany.

International trade will in turn affect the location of industry for example, the development of the Channel Tunnel from 1994 has led to more industry locating in Kent and Northern France. More recently, the Øresund Bridge and Tunnel which opened in 2015 links Denmark and Sweden and significantly reduces time.

The UK and Entry into the EU

At the time of British entry into the then Common Market in 1973, concern was often expressed that British entry would lead to development of the so-called golden triangle (i.e. that of South East England, Northern France and the then West Germany would gain as they were in the centre of the market). The reason for this assertion was that firms would tend to locate towards the centre of the market in order to minimise total distribution costs. This would be similar to the development of the West Midlands within the UK.

Sometimes domestic transport may be for longer distances than international journeys in general the longer distances involved in international transport means that there will be far greater emphasis upon speed and quality of service than for domestic movements. The referendum result in 2016 may alter the transport industry.

Common Transport Policy

One of the features, which the EU wishes to look at, is exhaust emissions. Clearly, there are problems, since some of the industry itself is very strongly opposed to standards, whereas most other people would wish to see a reduction in emissions partly because of problems of pollution and partly because of concerns about climate change.

Trans European Transport Network

There have been a number of concerns about this. One of them is that most of the investment in the rail sector is in the longer distance high-speed trains. This may be desirable if it means that there is a transfer from air to rail, but if it merely generates traffic, this is not necessarily an environmental advantage given the energy costs involved in creating the networks in the first place. In general, the people using the longer distance high-speed links are likely to be wealthy and we may see a repeat of the position in the UK that transport subsidies mainly benefit the rich rather than the poor. The same criticisms can be partially applied to the construction of new roads.

It is noticeable under the charging for infrastructure that the commission suggests that we need to look at not only covering the cost of the infrastructure whether it be road or rail, etc., but also the external costs which can be related to air pollution, climate change, noise, accidents and congestion and this should be across all modes of transport whether they are for private or commercial purposes. It also states that users should know what they are paying for and why. This is often referred to as transparency. Therefore, it whether transport infrastructure is used to improve safety or to minimise environmental disturbance, it should be clear who pays and also that the "polluter pays" principle should be endorsed. Most transport economists would probably agree to this principle, although they may well disagree as to what these costs actually are in monetary terms.

What Are the Obstacles to International Transport?

The Way

The way means the railway track, road system, sea or air through which vehicles travel. For example, the Berne Convention of 1890 occurred since if different countries had different railway gauges (i.e. the distance between the tracks), this would make international rail freight and rail passenger traffic more difficult. This would be similar to the decision eventually to scrap broad gauge railways within the UK for the same reasons. The standard gauge of

4ft 8½ inches has meant that trains for both passengers and freight can cover a great deal of the continental and British network without the expense or inconvenience for passengers and freight of changing at the borders. There have been problems for example with trains going to Russia (the former Soviet Union) and also to Spain. There are extremely expensive methods of overcoming this problem (e.g. adapting the locomotives or rolling stock). However, whilst the standard gauge is similar throughout the EU, the loading gauge is not the same. The loading gauge refers to the maximum height or width which rolling stock or locomotives can have. The loading gauge is determined by the height of bridges or tunnels or the distances between the separate tracks or between the tracks and platforms. In particular the Southern Region of the former British Railways has a generally smaller gauge than most of the continental system and also it is smaller than that of much of the remainder of the British Railway network. This poses problems when the Channel Tunnel allows the use of through trains. At the present time, it still poses problems since it means that the Freightliner network has not been extended to ports such as Dover in the South East and also that some European rolling stock cannot enter the UK at the present time. The increasing use of electrification partly because of lower maintenance costs and because of the expected higher costs of oil can also pose problems.

For example, since the Channel Tunnel has been built there is the third rail network of most of the Southern Region the use of the overhead pantograph system in most other parts of the UK and the overhead system used by France and the continental countries. This can be overcome by the use of special locomotives which can run on any system though this is expensive. An alternative which has occasionally been used in the UK is to have an electric system which is both overhead and third rail.

Apart from the individual characteristics of 'The Way", the railway network has generally been geared to national rather than international journeys. This can be seen in the comparatively slow speeds on many international journeys compared with the faster times currently for the London to Glasgow and in the future London to Edinburgh routes or even more impressively the Paris to Lyons route. Partly for this reason, therefore the Union Internationale Chemin de Fer (UIC) in the early 1980s suggested a master plan to speed up journeys between the main European cities. There is already a system of Trans

Europ Express System (TEE). Whilst the development of high-speed passenger rail links has perhaps significant prestige attached to it the less glamorous rail freight network is of potentially even greater importance. At the present time journey times between Britain and Italy for example can be quite long. There has been some co-operation between some of the Western European countries in developing the Hermes computerised system which would improve utilisation of railway locomotives as well as improving information to customers.

The Sea and Shipping Problems

For shipping the way is free, but there may be problems, for example because there is no obvious system of priorities for ships whilst at sea. At one stage for example within the English Channel the British and the French Authorities required one system whilst The International Maritime Organisation (IMO) required another.

The importance of this can hardly be overstated when it is realised that large ships may take up to 10 Kilometres (6.2 miles) to stop. There have therefore been considerable efforts by The IMO for the English Channel and also other crowded shipping areas to determine a system which would be much safer. The growth of international trade and in particular the super tankers has also posed problems where ships go outside conventional shipping routes or where the draught of the ships is much greater than previously. This therefore requires a system of charting and of international maps. The IMO and the International Hydrographic Organisation have co-operated in trying to improve ship owners' information. The provision of navigational aids can be regarded as a public good to use the economist's jargon. The economist would regard these as collective goods since they are not used up by more ships using these services.

There are also restrictions on the shipping because of the size of canals such as the Panama and Suez. The Suez Canal closure in 1967 was one of the reasons for the development of the super tankers. Once the Suez Canal was closed it was realised that in spite of the longer distances involved the economies of scale of the super tankers outweighed the disadvantages. In the late 1970s and early 1980s, the Suez Canal was widened so that it could take vessels of much greater dimensions. Because of the political volatility of the region there have been suggestions from time to time that an alternative canal should be built.

The Way and Air Transport

For air transport the way itself is free though there is an obvious need for air navigational aids to be available to all airlines. There is then the problem of attempting to charge for this and the problems which occur if particular countries do not have an adequate system. There have been from time to time complaints about the Spanish air control system. There have also been problems of capacity especially over British Airspace. In 1988, for example The Civil Aviation Authority suggested that there would be maximum limits imposed upon the number of aircraft flying over Southern England. This would have a particular impact upon the tourist business since the peak number of aircraft was likely to occur in the summer.

It would be difficult though not absolutely impossible for maritime countries to have a system of air transport without overflying other countries. With East and West Pakistan, up to 1971 even domestic transport could not have taken place without overflying a third country (i.e. India). The problem is now an international transport since the formation of Bangladesh following the civil war. The Chicago conference in 1944 therefore proposed five freedoms, the first freedom was the privilege to cross territory without landing; the second was to land for non-traffic purposes; the third was to be able to put down passengers, mail and cargo in the territory of the state for the particular nationality of aircraft, for example a British aircraft could put down passengers, mail and cargo within the UK; the fourth privilege would be the privilege to take on passengers, mail and cargo for the territory of the state corresponding to the nationality of the aircraft, for example a British Aircraft could take on passengers, mail and cargo to put them down in Britain; and the fifth privilege was the privilege to take on passengers, mail and cargo destined for the territory of any other state which had accepted the agreement and also to put down passengers, mail and cargo from any such territory.

The Need for International Road Routes

The road network has fewer obvious problems with the way than other modes of transport. However, the road network as with other modes has generally been geared to a network serving domestic rather than international needs. Partly for this reason the AGR Convention 1975 led to the provision of a number of 'E' routes (European routes). The British government has not so far ratified the convention in this respect. The AGR Convention lays down standards for road widths, sections between junctions, intersections and so on.

An example of an 'E' road is one which goes from Cork in the Republic of Ireland via Fishguard, Bristol, London, Felixstowe and then on via the Hook of Holland to Warsaw and Moscow. At the present time 'E' routes have not been marked in the British Isles.

There have been considerable delays at frontiers for both road passenger and road freight traffic. The EU Directive 72/166 abolished the checks on Green Cards at frontiers within EU states so that they should have reduced delays.

More serious delay has been the checking of loads at frontiers. There has therefore been a Transports Internationaux Routiers (TIR) system under two conventions in 1959 and 1975. Under the TIR Regulations providing the vehicles move in sealed secure labels with the TIR system, they will only be inspected at either an inland clearance depot or the first frontier and the last frontier or the inland clearance depot of their destination. There may still be random checks. Within the EU TIR carnets do not apply, since there is a community transit procedure for EU members as well as for Switzerland. This has much the same effect.

Road hauliers would find it difficult to carry on international road freight if they were not allowed to go to another country without having paid the vehicle excise duty of that country. Under the International Convention of the Taxation of Road Vehicles to which all the EU member states belong as well as a number of other countries there are no fiscal charges on UK vehicles entering those countries or of course for vehicles from those countries entering the EU.

There are, however, still taxes imposed by certain countries particularly if they traverse the country without carrying goods to or from that country. Austria levies a charge for each tonne of permitted payload capacity per kilometre. This is partly because Austria is one of the countries which has many international lorries traversing it without it being of any immediate benefit to the Austrians. Partly for this reason therefore there have been suggestions from time to time, particularly since Greece joined the EU in 1981 that the EU might make some financial contribution towards the main road network in Austria since otherwise this is an obstacle to EU trade.

Road hauliers would also find it very difficult to carry on their trade if they had to comply with the detailed construction and use regulations of other countries, for example France has regulations which specify the use of yellow headlights. Therefore, there have been type approval provisions whereby if a vehicle is accepted by one country within the EU, it has to be accepted by the others.

Documentation

This may be an obstacle to international transport partly because of the costs filling in such forms and partly in some cases because of the deliberate red tape for some countries. Italy for example has often been accused of being excessively bureaucratic. International organisations such as the IMO have therefore spent considerable time trying to facilitate documentation procedures. The International Road Union has similarly tried to improve the documentation for road hauliers. For airlines, The International Air Transport Association has a standard waybill which means that most consignments can be sent using this one document.

Within the EU documentation has been a barrier to trade and therefore from the 1st January 1988 the Single Approved Document (SAD) procedure has applied. The then Department of Trade also used a system called The Simplification of National Trade Procedures Board (SITPRO) which used a system of aligned documentation. SITPRO has now been abolished.

Wages

International transport companies will wish to obtain the cheapest labour which is compatible with giving the right quality of service. This poses a number of problems, for example American Shipping Companies if employing American employees will generally have much higher costs of labour than if they recruited labour from elsewhere. There have therefore been regulations preventing American Shipping Companies having imported labour. This in turn has meant that many American Shipping Companies have gone under flags of convenience (i.e. sailing under the flags of another country in order to gain both this advantage and also frequently tax advantages). In the UK, British Shipping Companies have often put their ships under flags of convenience for example Cunard's have done this so that they too can obtain cheaper labour. The British Maritime Fleet has declined drastically partly because of this.

Demand for International Transport for Tourist Reasons

The tourist trade has grown rapidly in recent years. The UK has over 3 times as many UK residents making trips to other countries, even though the UK is also a tourist country. There have been over 2.5 times the volume of visits since 1980s by overseas residents visiting the UK.

The increasing length of holidays plus cheap flights especially with the so-called budget carriers have accentuated this trend.

Whereas in the 1960s there were often restrictions on the amount of money that people could take on their holidays with them, the lifting of any foreign exchange controls will have meant that people now go more often and on more expensive holidays than where such controls were in force.

Visiting friends and relatives (VFR) is also important and many people now also have second homes abroad. However, timeshare which was a rapidly growing industry has faded slightly in recent years.

Different Patterns of Freight

Where traditionally the UK both imported and exported goods mainly with the Commonwealth this has become less important and more trade has been with the other 27 countries of the EU. This has implications for the UK port industries since the East coast ports particularly Felixstowe have become more important whilst the West coast ports have suffered a decline. As the ships get bigger for economies of scale reasons the ports have sometime changed their location with the Port of London moving to Tilbury rather than the traditional East End and more recently Thames Haven in Essex and Thamesport on the Isle of Grain in Kent has further reduced Tilbury's importance.

CHAPTER 25

Developing Countries

It has become more difficult to define developing countries especially since the Organization of the Petroleum Exporting Countries (OPEC) price rises in 1973-1974. It is perhaps easier to recognise developing countries such as India, Bangladesh and Pakistan rather than to define them. In general, economists think of developing countries as having almost by definition a low income. Some economists, however, distinguish between the newly industrialised countries (NIC) which have fairly high economic growth rates and other developing countries.

There is not an adequate definition of a developing country. Most definitions have been based on the Gross National Product (GNP) or Net National Product (NNP) per head. The problem with this measure is that it is more useful where people have similar patterns of lifestyle and expenditure. It is not very useful when economists try to compare a subsistence style economy with one which is specialist. The use of GNP and NNP only refers to absolute levels of incomes measured in terms of a given international currency for example the dollar. This method of measurement has understated the levels of income and development in developing countries. A more useful method of comparing development in different countries is the 'Purchasing Power Parity' (PPP) approach as proposed by the World Bank. This measure seeks to capture the actual value gained from different currencies in different countries for a general basket of goods and services. For example, travelling a distance of 60 miles, for example from Canterbury to London will cost £30 whereas travelling the same distance in Kenya, say from Nairobi to Kenol would cost KES150. Assuming the exchange rate is KES130 to £1, so it would cost £1.15, then travelling the same distances in the UK is 26 times more expensive than in Kenya. This also means that a person with £30 in Kenya is better off than a person with £30 in the UK.

There have been suggestions by the Organisation for Economic Co-operation and Development (OECD) that we could have measures which look at standard of living more directly (e.g. mortality, health, access to housing, educational ability and so on). There have been changes in the patterns of developing countries particularly since 1973 when Oil Petroleum Export Countries (OPEC) raised its oil prices. Partly as a result of this, some Middle East countries became very rich. These measures are known as the Human Development Indicators and the Human Poverty Index which are used to show the standards of living they pay more attention to other aspects of a wholesome life apart from money. These aspects include political freedom, access to human rights, medical care, and safety from crime amongst others.

Human Development Index

The Human Development Index (HDI) was created by the United Nations (UN) created to emphasise that people and their capabilities should be the ultimate criteria for assessing the development of a country, not economic growth alone. The HDI can also be used to question national policy choices, asking how two countries with the same level of gross national income (GNI) per capita can end up with different human development outcomes. These contrasts can stimulate debate about government policy priorities.

HDI is a summary measure of average achievement in key dimensions of human development: a long and healthy life, being knowledgeable and have a decent standard of living. The HDI is the geometric mean of normalised indices for each of the three dimensions.

Skewed Income Distribution

Another problem is that the income distribution is skewed (i.e. that in many countries there are a few very rich people, whereas the many people have a very low income so that an average figure is not typical). In recent years as well, there has been an increase particularly in the 1990s NICs, such as those in the so-called Tiger States of Southeast Asia. More recently, both China and India have had very high rates of economic growth, although in both countries there is still considerable poverty as well as some very rich people. The developing countries usually have both high rates of unemployment as well as underemployment. Underemployment occurs particularly with the extended family when more people are employed than would be necessary to carry out the allocated tasks. The term extended family is used to mean where the family consists of more than the fairly typical nuclear family in the west, of husband, wife or partner and a small number of children. The extended family by contrast may consist of three generations as well as cousins, uncles and aunts. The extended family may alter the demand for transport, since people will not have to travel far to visit relatives and will also alter the supply since often the extended family may run small scale transport undertakings.

The Gini coefficient is a measure of just how skewed income distribution in the countries. It is a ratio between 0 and 1 where 0 represents perfect income distribution and 1 represents a situation which at times may have been witnessed in Sub-Saharan Africa where the majority of the productive resources belong to a few individuals.

Lack of Skilled Workers

In many cases, there is a lack of skilled workers who could carry out adequate maintenance. Lack of skilled labour has been exacerbated by the lack of opportunities for the skilled workers who opt to migrate to countries where their skills can be absorbed into the market. This is known as the 'brain-drain'. In some countries, constant political instability and even death of experts may lead to a situation where skills cannot be fostered.

Some of the African countries such as Kenya, South Africa and Nigeria have developed powerful institutions of higher learning as well as collaboration with universities abroad to provide education. African countries also face a debilitating haemorrhage of skilled workers to developed countries. The recent drive to recruit care workers from developing countries by the UK attracted young, ambitious and highly educated nurses and medical staff from the poor countries of Africa and South Asia.

Poor Infrastructure

Poorly built and maintained roads are also a serious problem in Africa. This is one of the reasons why vehicles in some of the poorer countries often have a much shorter commercial life than would be common in a country such as Britain with its railway systems. The poor infrastructure also does not help whether it is the roads or the rail system. There are relatively few all-season roads outside towns as a good atlas showing Africa will show you. The road system needs to be improved rapidly in order to cater for the increase in population particularly in most urban areas. In the rural areas, there is a need for some roads if only to get agricultural produce to and from these areas. In some cases, the second milking of the cows is wasted.

Human Resources Management

This is often poor though this criticism is not confined to developing countries. It is often compounded in developing countries where people may be chosen because of their tribal or clan grouping, political affiliation, or because they are part of the family. The concept of meritocracy has its shortcomings but usually leads to better management. Three is a need to look at overall manpower planning when looking at the future needs of many industries including transport and frequently this has not been done where promotion is on grounds of favouritism or seniority rather than ability. There may be a failure to consider modern management methods and instead to rely on old procedures. Lack of knowledge of operational research techniques such as queuing theory, network analysis is often poor although again this is not just true of developing countries.

Lack of Standard Gauge Railways Compared with Europe

The railways in many cases were built to transport minerals to the coast as in West Africa. There was little need to consider other facilities and in many cases the railways are not of the standard gauge of 4 ft. 8½ inches. This makes it more expensive generally to obtain rolling stock or locomotives, since there are not the economies of scale which would be available in Western Europe.

Railway Systems in Indian Subcontinent

In some countries such as India, there was a widely developed railway system which has the advantage of providing for very large numbers of passengers per hour. This is much cheaper than the same number of people in cars or motorways; however, surface railways do not generally penetrate into the centres of most existing cities. If separate land is required for tunnelling or viaducts, then fixed costs are likely to be high.

Underground Systems

Underground systems have the advantage of using little land space. They can be sub-divided into two types, either cut and cover where as the name implies the land is cut into and then covered over as for example the Metropolitan, District and Circle lines in London are. This is generally cheaper than the Tube where as the name implies tunnels are made as with the Piccadilly and Northern lines. Tunnelling is usually expensive and will only be justified if land space is very expensive as with Hong Kong and Singapore.

Steam Railways

In some countries steam railways still exist. Steam Locomotives have the advantage of very long lives and do not usually break down. Higher speeds which are obtainable with diesel and electric trains are often less important than in richer countries. However, most countries have signed up to climate change conventions and this will not help.

Need for International Railway Routes

There is the need to consider railways for international as well as domestic transport. Nigeria whilst a very large country has no international railway routes whereas the Tanzam railway was built almost entirely for international reasons.

Ports Development in Poorer Countries

With ports the more modern methods of handling such as containerships have been less helpful to developing countries, since it requires expensive capital equipment and less labour which is the opposite of what these countries require. On the other hand, roll-on/roll off ferries and packaged timber are useful. The use of computers for data may be helpful since documentation has often been one of the major causes of poor utilisation of ships in ports.

Port development has often become more expensive because of the need to cope with containerisation handling equipment such as straddle carriers which are expensive. It has the disadvantage of being capital intensive. Some developing countries have had fairly large ports and fleets partly for balance of payments purposes.

Decision Making by the Rich

Since the rich usually make the decisions about transport, facilities for pedestrians and cyclists are often neglected in favour of more prestigious methods of transport. Cycling has the advantage that it only needs a minimal width compared with conventional roads.

Whilst cycles are usually thought of as a passenger vehicle the North Vietnamese showed in the civil war in the early 1970s that loads of up to 300 lbs (140 kg) could be carried. The use of separate cycle ways would open up more employment opportunities to the poor since they would then be able to take longer distances to and from work. At the present time, poor standards of road maintenance as well as poor driving standards means that many cyclists are deterred. A World Bank report on urban transport implied that cycle ways are an efficient use of investment.

Cycle Tracks

Cycling is non-polluting and even if battery operated cycles (electric bicycles) are used there is little pollution or noise. Cycle maintenance is cheap and can lead to self-sufficiency compared with imports of expensive cars. Additionally, much less land is used for parking space. The need for a low-cost public transport system or better provision for cyclists can hardly be overestimated.

Alterative Vehicles

Even bus fares which are low by Western European standards account for a large percentage of low incomes. There is a need to develop low cost vehicles which are mechanically reliable and can take a reasonable number of passengers even if standards of comfort are low. Care should, however, be taken that overcrowding of vehicles does not lead to greater accident risks.

Use of Intermediate Size Vehicles

The use of intermediate size vehicles with perhaps 12 to 17 seats which are variously called Matatus, jitneys or jeepneys (Figure 25.1) may be helpful, these have low capital costs compared with conventional buses and may be able to reach parts of town which larger buses could not reach. Sometimes they may be also used to carry freight. They are often run as a small family business. The use of maxi taxis may also be helpful. This implies that there is a kind of door to door service which is cheaper than with a conventional taxi. This is not likely to be helpful to the very poorest people but does relieve some of the acute congestion in many urban areas.

Figure 25.1 Jeepneys – an informal public transport in the Philippines (Photo: Mengqiu Cao).

The use of small vans had been popular for a long time in Kenya. The name 'matatu' was coined in the early 1950s when vehicles for carrying paying passengers were introduced then the price of travelling anywhere was 30 pence (translated in Bantu to 'mang'otore matatu') (i.e. 3 × 10 pence coins). Over the years, the vans became smaller and carried less people. It was not until early 2000, when the economic growth hit a high of 5.8% from a previous low of 0.5% that the place of matatus has had to be revaluated. Characteristically manned by loud unruly youths, the matatu business was faced with increased private transport and a government bent on resolving the unruliness once and for all. The Kenyan government has now issued a directive that vehicles that seat less than 18 passengers are to be phased out in favour of buses that carry over 46 passengers. This is to ease the congestion in the city. With time the Kenyan government intends to encourage the use of larger buses including the 'bendy-bus' to ease congestion and make transportation safer in the cities.

Congestion and Causes

Whilst car ownership is lower than in developing countries the problems of urban congestion are often acute partly because of lack of adequate roads and partly because the traffic ranges from animals to modern cars. Traffic lights rarely exist and policemen seem incapable of keeping control. Subways or footbridges are rare partly because of relatively high capital costs. However, the need for adequate pavements is important and more pedestrian crossings may offer scope for enforcement of safety. There is unlikely to be the scope for large scale subsidies to public transport which is found in many parts of Western Europe.

Priorities for public transport such as bus roads or bus ways could be developed at relatively small costs. The advantage of a segregated busway is that it is much cheaper than an urban railway system. Other fixed costs are lower than for railways or tramways. Intervals between buses usually referred to as headways are lower than with conventional rail systems.

Tramways with a segregated system might be helpful, if there are potentially cheap sources of fuel available such as hydro-electricity. They, however, have not generally been used because of the high fixed costs. As pollution is often a major problem the use of either electric cars or a due powered system which might use batteries as well as overhead cables could help the two overcome the problem of serving routes in the middle of town.

Need for Town Planning

There is often very poor planning with the needs of the shanty towns often being ignored. The rapid change in the size and population of the urban areas often means that planning and transport should go hand in hand with provision of cycleways, busways and possibly even roads which could be used in one direction in the morning and the reverse direction in the evening.

Air Transport

For inter urban transport where the terrain is difficult as in Botswana the use of light aircraft and air strips may avoid the high costs of infrastructure. It may also help development in towns which are in otherwise rural areas. Development of airstrips may be helpful for urgent consignments such as medicines and the occasional passenger such as medical staff who may have to travel regularly and urgently.

Transport in Rural Areas

Income in rural areas are usually very low, because paved roads are comparatively rare as can be seen from many standard atlases, large numbers of the population may live at considerable distances from the nearest paved roads. One alternative to building more roads would be to use vehicles such as Land Rovers which can use low cost tracks. Cycle tracks being narrow are cheap to construct, there is little need to have roads built to European standards. Sometimes it would be better to redesign vehicles to suit the roads. It may be possible to adapt and improve the use of animal power. This can be done with bullock carts. Additionally, motor cycles or cycles with trailers might be helpful since the track width is lower than for cars. In many cases, there is a need to develop vehicles such as buses which can be used for both freight and passengers. It would be helpful if the vehicles could be designed or at least manufactured within the country rather than relying on imported technology. Imported cars cause problems when breakdowns occur, making services unreliable.

Alternatives to Transport

There has been over emphasis on mobility rather than accessibility. In some cases, the use of distance learning for education might be cheaper than to have teachers going to the rural areas. In some cases, boarding schools have been used to avoid the problems of children having to walk very long distances to and from school.

Transport and Economic Growth

The World Bank has often supplied cash for infrastructure, since it is hoped that it leads to economic development. Many developing countries have invested in transport for the same reasons. It is important that before transport investment takes place, it does not replace activities elsewhere in the economy for example, a new highway may stimulate production in one part of the country but merely replace production elsewhere. There is little point in being able to import American rice when it will be better to consume home grown rice. We need to consider what happens with and without transport facilities to ensure that the returns are greater than from the provision of pure water supplies or medical facilities.

Population Growth and Difficulties in Estimating Population

In most developing countries, there is a much higher increase in population than in developed countries. This means that changes in population affect both the land use and the modal splits more quickly than would occur in most developed countries. In the majority of developing countries, there is also an urban drift as though incomes in towns may be low, there is a greater possibility of jobs than in agricultural areas where there is less likely to be any increase in availability of jobs.

Many developing countries will have a different age structure to those of developed countries. The proportion of people under fifteen is much higher than in a developed country and therefore even though the demand for school services may be in some respects lower (i.e. there is less likely to be further or higher education than in a developed country, the total proportion of journeys to and from school is likely to be greater than in a developed country). This is in contrast to the UK where for the first time we have more people over 65 than under 16.

It is difficult to obtain adequate data for many developing countries which is in anyway accurate. Even the population estimates of a major country such as Nigeria in the 1980s varied between about 80 to 120 million. If the population is difficult to ascertain, it becomes even more difficult generally to obtain adequate information about passenger or freight journeys, although this problem is by no means confined to developing countries. In Federal countries estimates are often biased since the amount of money spent as well as taxation may depend upon the estimated size of the population within the different component states.

Road Congestion

Whilst incomes may be low, this does not mean that there is no problem of road congestion. This partly arises because as in many other countries the road system was not geared towards modern traffic. The problems are perhaps even more acute, because there is a wider range of road users ranging from animal traffic to the latest cars. The World Bank has suggested in one of its reports on urban area in developing countries that in many cases traffic control is also inadequate. This problem again is not, however, unique to developing countries as anybody who has visited a city such as Rome will be aware. One of the possibilities which economists including those from the World Bank have suggested is that there might be greater priority for public transport. In developing as well as developed countries there is a need for a more rational use of road space which might not only relate to physical planning which creates the need for transport in the first place but also to try to ensure that the existing system takes the maximum amount of traffic.

Importance of Buses

In most developing towns, the bus will be the main form of transport though there are some exceptions such as tramways in parts of Asia and South America. However, the typical life of such vehicles may well be less than in developed countries partly because of the problems of poor maintenance and partly because road surfaces often leave a lot to be desired. As an illustration of this point one Polish professor came up with an estimate that in Kinshasa in the 1970s the typical life of a bus was about three to four years even though the number of workers per bus was much higher than in a typical developed country such as the UK.

Restrictions on Car Ownership

One possibility which would make better use of the roads would be to restrict car ownership either by higher taxes or quotas. Particularly for countries which have a balance of payments problem; restrictions on imports to cars might be helpful. A quota system might be unfair and in any case liable to evasion the higher taxation system perhaps reflecting the shadow costs of foreign exchange might well be much more economically desirable and practical. Nigeria at one stage had a system of only allowing different cars on different day in some areas. As with developed countries there might be a need for staggering of working hours especially in those cities where there are a limited number of possible access roads and also better traffic management schemes, for example as elsewhere the use of one way streets parking restrictions. This would generally be cheaper than to extend capacity through the use of new roads. Because cycling and motor cycling is much more common even in fairly wealthy countries such as Singapore than in most developed countries there may be a need for bus lanes and also for separate provision of tracks or roads for cycles and motorcycles. The use of (para transit) may be helpful. These are often run as in Uganda by families. The capital required for them is lower than for conventional buses and may be within reach of a richer family. Family ownership may also mean that problems of the peak can be overcome more easily than by more conventional public transport. The use of shared taxis is common and is also helpful. It seems from the likely evidence that the number of taxis is much higher proportionately in most developing countries than in developed countries.

Pedestrian Safety

There is also the need to ensure that pedestrians have greater segregation and therefore safety. The World Bank in particular has suggested that they would prefer to provide finance for such concepts rather than on commercial traffic schemes basically designed for car users.

There is perhaps less scope for staggering of hours than in some other countries since press and other reports suggest that Bangkok in Thailand and Lagos the former capital of Nigeria may have congestion problems for twelve hours a day compared with a typical four or five a day in many developed countries.

Car Ownership and Economic Growth

Whilst it is often claimed that there is a link between car ownership and economic growth there is no reason to believe that car ownership causes economic growth one might assume more easily the converse that economic growth leads to greater demands for car travel (i.e. as car travel is often a consumer rather than an investment decision). Countries such as the People's Republic of China, Korea and Columbia all have had low rates of car ownership but relatively higher rates of economic growth. Part of the reason for disparities in car ownership with similar levels of income is the vast differences in taxation on petrol.

Cycling

Whilst cycling is usually only considered as a method of passenger transport, it has been used quite often for freight. This was dramatically shown during the Vietnamese Civil War in the 1970s when the North Vietnamese in particular often used cycles for carrying very large amounts of goods. One of the problems for developing this concept is that most cycles have been designed in the Western European countries for personal rather than freight transport. The advantage of cycles is not only that they are cheap, but also that they are non-polluting and that the amount of track which needs to be provided is much smaller than for most motor vehicles with the exception of motor cycles. The Intermediate Technology Group would go much further than this and suggest that there should be much more emphasis given to the development of traditional animal transport including the ox cart. This is increasingly important since the rise in oil prices in 1973–1974.

Fares Policy

Because the poor are unlikely to be able to afford any type of public transport which involves any great distances, for example a typical fare for two or three miles (four to five kilometres) per day would often amount to about 10% of an average worker's income, it may be sensible to have much higher densities for land use planning than would be common in Western style countries. This could make public transport more viable.

Building Vehicles within Countries

One possibility to try to avoid balance of payments problems is for a much greater emphasis on building vehicles within countries. Where countries are too small to provide their own vehicle manufacturing facilities it might be necessary to have some form of regional grouping (e.g. Economic Community of West African States rather than for individual countries to try to form their own manufacturing). This would be similar to developments within the EU.

Supply of Transport

Whilst economists have tended to look at demand factors, it is also important to look at the supply of transport. Where the use of roads cannot be improved or increased it may be sensible to look at ideas such as those of a light railway which would involve lower infrastructure costs than traditional railways. Concepts such as those of the Docklands Railway in London may not be entirely sensible, since there is less need to have small amounts of labour because of the lower wages involved. The basic concept, however, of being able to go around tight curves and up steeper gradients may well however be beneficial if the train is otherwise unsuitable for rail transport. It may be sensible in many cases to try to improve manpower planning within many of the larger public corporations in particular rather than spending more money on infrastructure.

This would certainly seem to be true, for example of some of the Nigerian Public Corporations. As elsewhere, there is little point in trying to improve vehicle speeds particularly for freight if documentation procedures do not also improve. There were many complaints for example in the 1970s about the slow documentation in Tin Can Island (the port for Lagos). The use of modern technology such as computers, fax or emails means that there is considerable scope for improving documentation procedures without too much investment capital being needed.

An improvement in manpower planning as well as a greater use of management techniques including the use of computers might help in the use of preventive maintenance as well as improving fuel consumption through the use of operational research techniques.

Sometimes countries have gone for the use of outside consultants, for example Nigerian Airways in the early 1980s used KLM the Dutch Airline for advice.

Transport expenditure is of course not the only priority for developing countries, for example preventive medicine and provisions of pure water would save large numbers of lives. However, the use of transport facilities will be helpful providing it is linked through either administrative or economic measures to some form of national planning.

A small country such as the Seychelles has shown that there can be links between the national development plan and transport. The Seychelles National Development Plan in the 1980s looked at the role of Civil Aviation which is particularly important because of the importance of encouraging tourism as well as providing links between the different islands and also using it as a crossroad for inter-continental air traffic. The Seychelles has also wanted to develop its port and maritime services to meet the Seychelles demand for both imports and exports.

Other countries such as Botswana a landlocked country in Southern Africa have also tried to link their national planning with transport and communications planning.

Whilst in countries such as the UK the use of metalled roads is taken for granted this is by no means the norm in many developing countries. Therefore, there is perhaps a need to develop off the road vehicles perhaps broadly similar to the Land Rover. It may also be sensible to look at alternative sources of fuel for example the use of alcohol, ethanol or even sugar. Ethanol is now being grown extensively in Brazil for this purpose.

In many cases countries may be too small to do their own training. Therefore, it may be helpful if countries such as those of the front line states including Zambia can come together with other developing countries in developing joint training. In other cases, there may be a need for international co-operation for example dealing with the problems of shipping such as the 40-40-20 proposals.

Subsidies for Public Transport

There is often a justification on social grounds for carrying certain categories of people for example school children below the cost of running such services. However, even more in developing countries than in developed countries there should be a transfer from general funds to the undertaking to show the effect on both costs and revenue of such subsidisation. In general, there may be a case for subsidies if congestion taxation is not being charged though there is a much stronger case to have congestion taxation. Subsidies therefore are very much a second best solution to use the economists' picturesque phrase and they distort competition between modes of transport. If the subsidies lead to low fares for longer journeys this will mean that over great provision is being made to the transport sector compared with other urgent priorities including primary health care. However, subsidies which are well thought out may in the short run be a much more sensible approach to urban transport than greater road provision.

Co-ordination Between Transport and Planning

In many countries, there are often squatter areas though the names for these would vary between countries. These often have no bus service partly because countries do not want to recognise such squatters and therefore have not built adequate roads.

International Collective Agreements

Developing' countries might be able to reduce their balance of payments problems through collective agreements. For example, the United Nations Conference on Trade Aid and Development (UNCTAD) suggested a forty, forty and twenty roles for liner conferences. This means that 40% of the trade between two countries would take place in the importing countries ships 40% in the exporting country and only 20% of trade will be left to so-called cross traders. This action was suggested because developed countries notably Britain in the past, as well as Eastern Europe have a large part of the total shipping market. The move, however, would be against the principles of comparative advantage and therefore could lead to higher charges for both import and export. There might be the need to lower charges, since otherwise the consumer would presumably have chosen ships from their countries in the first place. However, the infant industry argument could be used that if developing countries managed to succeed in the short run then the need for protection might be less true in the longer term.

There could be protection to local shipping companies. In the short run, at least, however, it would add to both import costs and export costs. How much difference this would make would obviously depend upon the cost structure of the developing countries compared with the developed countries. One would need to weigh up this disadvantage against the potential balance of payments savings.

Even this, however, is fairly complex, since most ships would have to be bought from developed countries and thus in the short run the developing countries would have to pay money which would be shown in the capital accounts of the balance of payments. Similarly, many developed countries including the UK use labour from developing countries noticeably from the Indian sub-continent. Therefore, some of the current money paid for example to British ship-owners will find its way into the pockets of developing countries citizens. A further problem is that of the poorest landlocked countries, for example Zambia might have to pay higher charges without any offsetting benefits unless they can in some way join the shipping countries.

The other possibility which needs to be considered is whether in some cases if shipping conferences did not serve developing countries ports would there in fact be a service at all. On the other hand, the developing countries may well feel that shipping conferences were a form of cartel which tend to operate by restricting supply especially with closed conferences and raising prices though the shipping conferences would deny this. However, both the EU commission disliked conferences because of the Treaty of Rome competition rules and the Americans whilst originally giving conferences exemption from monopoly legislation by the US Shipping Act had tended also to favour more competition. However, shipping conferences have now been abolished so that discussions no longer matter.

Transport in West Africa

In West Africa, as in some other countries the first forms of transport development were near the coast partly because of the lack of suitability of suitable rivers. These were often formed because of the need to trade with Europe and so external trade was developed before large scale internal trade. Later there were developments of transport inland partially to obtain mineral deposits. This was true for example of Sierra Leone. There is also a need for inland transport so that agricultural products could be exported. Another reason for development of transport was to have some form of government control over larger areas.

In Nigeria, as in many other African countries the rail network is extremely limited with railways running to the two main ports of Lagos and Port Harcourt. The line to Port Harcourt was originally developed mainly for mineral traffic. Whereas British Rail was reducing its network considerably in the 1960s, this was not true in Nigeria where one major new line of over 600 Kilometres was being built. The speeds on such railway services were much slower however than in Europe, for example the line from Southern to Northern Nigeria took over 30 hours to cover the journey in the early 1980s and was not necessarily very reliable. The Nigerian railways lost traffic for a considerable time during the 1970s and by 1978 were only loading 65 wagons per day and carrying about 800,000 tonnes of traffic a year. There had been moves as in other countries from steam to diesel, but the utilisation of many locomotives was very poor and breakdowns of certain types of locomotives were extremely common. The Nigerian railways therefore asked Rail India Technical and Economic Services for commercial advice. This seems to have succeeded so that by the early 1980s about 80% of the locomotives were being used and about three million tonnes of traffic were being transported. The passenger services were carrying almost double the number of people in 1981 compared with 1978. The main reason for these changes was not large-scale capital injection but making much better use of existing assets and in particular paying very strong attention to spare parts. In January 2007, the Federal government announced that it would try to modernise the railway network through a PPP model in a public private partnership. The estimated cost of the total project was over $40 billion. The aim is to have around 40 to 50 million tonnes of goods transported over the modernised network per year.

In East Africa, the three countries of Kenya, Uganda and Tanzania are trying to have an integrated transport system. Uganda and Kenya have appointed the Riff Valley Railway consortium to manage their railways over the next 25 years.

Difficulty in Describing Developing Countries Railways

Whereas some countries such as Botswana which had its railway corporation formed in 1967 when it gained independence from Britain and has a relatively sparse network which connects the border of Zimbabwe Francistown–Gaberone (the capital) to South Africa along 888 km of narrow gauge track as well as two branch lines, other countries such as India railways have a thriving network with currently both passengers and freight traffic increasing by over 10% per year.

CHAPTER 26
Recent Developments and Future of Transport by Mode

Developments in London – T-Charge

It is all part of the Mayor's plan to tackle air pollution. Around 9,000 early deaths in London are thought to be caused by long-term exposure to air pollution each year. As a result, Mr Sadiq Khan, the Mayor of London, is investing £875m over the next five years to alleviate the problem. The "T-charge" (officially known as the Emissions Surcharge) has been introduced since 23rd October 2017 and has been in operation Mondays to Fridays from 7am to 6pm. The T-Charge operates in the same area as the Congestion Charge Zone.

Other initiatives include the launch of the world's first "Ultra Low Emission Zone", which would mean that all vehicles, except taxis, had to adhere to minimum emission standards. In addition, it should be noted that the T-Charge will be replaced by the Ultra Low Emission Zone which will mean vehicles using central London will have to meet new, tighter emissions standards from 8th April 2019. This will affect all vehicles.

Encouraging Cycling

Some firms, notably Body Shop (a British cosmetics, skin care and perfume company), have tried to encourage their workers to come to work by bike, and some local authorities have adopted a scheme whereby cycles can be subsidised. The logic of this from the local authority's point of view is that car parking space is expensive to provide, whereas cycle sheds are cheap. There seems to be no reason why for short distances cycling allowances should not be at the same rates as cars. In order to encourage people to cycle, cycle lockers might need to be provided. Some countries notably Sweden have long distance buses, which can carry bikes so that in semi-rural areas that people can often use their bikes for relatively short distances, but leaving the bus to go along the main road for longer distances.

There has been considerable controversy about carrying bikes on trains. The rail operators have suggested that bikes take up room, but since in very few cases (i.e. only in the peak is the capacity fully used the marginal cost of providing for bikes is usually negligible).

Central Government Could Take Lead

There seems little point in the government saying it is concerned about climate change if both senior MPs as well as senior Civil Servants do not take the lead in trying to reduce their own carbon footprint.

Taxis

The advantage of taxis is that they use fewer parking spaces than if all the people who use them in the centre of towns were to come in by car. There seems to be no reason why hybrid vehicles (use of battery and diesel engine or petrol) could not be used by taxi operators. In the London area, taxis have priority over other traffic.

Electrification of the Railways

There has been concern that very little apart from the extension of the Channel route to St. Pancras in November 2007 and one new electrification project as a result of the improved West Coast main line system had taken place with the Labour governments from 1997–2010. In Scotland, the new railway link from Airdrie to Bathgate will be electrified. During the coalition government 2010 to 2015, more electrification took place including much of the route from Paddington to Bristol, Cardiff and Swansea.

In 1981, the British Railways Board and Department for Transport (DfT) had issued a joint report on electrification. In spite of the suggestion of a rolling programme of electrification which would have meant that the same team could have been used for an overall system very little has been done. The report was written at a time when the assumption was that there would be increases in the intercity as well as the urban freight movements. In 2004, only 31% of the total route network in the UK was electrified compared with 52% for West and Central mainland Europe. Of the 100 largest cities in the EU in

2006 only 15 were not served by electrified main lines and 5 of these were in Great Britain.

It is very noticeable that from London to Aberdeen that diesels are being used for the whole distance, even though from London to Edinburgh the line is electrified. Therefore, there have been suggestions that there might be hybrid trains which could use electricity where the lines are electrified and other forms of fuel elsewhere.

If we are to avoid the problems of climate change, we would probably need to have more densely inhabited cities and towns, rather than more urban sprawl. If this were to take place, then it would make rail services more attractive and also electrified rail services more attractive as well. It is very noticeable that in 2007, Virgin was spending considerable amounts of money on advertising its services from London to Manchester and that it also mentions very strongly the much better fuel use than with roads. It seems very strange that whilst there have been protests about the reduction in Eurostar services from Ashford that little attention has been paid to the poor quality of service from Ashford to the South Coast although in fairness there have been improved services in the last couple of years. If this line were to be electrified from Ashford to Hastings then it would seem much better to do this and to ensure that people could get quickly from the South Coast, especially as it is now more difficult to get from this area than when Waterloo was the terminal station.

We might also need other measures so that regenerative breaking was automatically used and also possibly solar cells.

The DfT has come under attack for not carrying out more electrification. The White Paper on the railways acknowledges both that electric traction is more 18% more efficient than diesel and also gives a greater carrying power than diesel trains. At the present time the total fuel cost for electric traction is around £225 million and rail travel only represent about 1% of total carbon dioxide emissions, whereas transport as a whole is responsible for 23%. It seems unlikely that hydrogen will be available as a fuel for a considerable time and the advantage of electrification is that it can use a wide variety of fuels. This would seem particularly helpful when the North Sea oil runs down that we are going to be reliant on unstable regimes in the Middle East and elsewhere which does not seem to be very sensible.

Currently, there is to be a new fast link to Bristol and South Wales, since the traffic to and from Paddington is rapidly approaching capacity and even though Reading Station is to be expanded, there will still be some problems. Whilst high speeds use more fuels, they would still use less than aircraft. It is hoped as well that rail freight will continue to expand and so by having separate track compared with a high-speed link that there would not be capacity problems with this. For the journeys to and from Wales the traffic on the M4 could be reduced, and this would be helpful in reducing both fuel and other social costs.

There is less acceleration and deceleration so that the overall fuel use may be no greater than with slower trains in spite of the higher speeds.

High Speed Link from England to Scotland

There has been considerable discussion about the possibility of a new high-speed link, from England to Scotland partly because it is difficult to see how the capacity of the railway system can grow without such a development. One of the advantages of such a system would be that at the present time the majority of long distance traffic goes by air, whereas if the travel times between London and Scotland could be reduced then this would have a significant effect on the demand and this would be good for the environment. It is very noticeable in other countries noticeably France that the TGV network has increased significantly and has been financially successful as well. Many commentators would also like it to serve Heathrow so that where there is little choice as for example with the Atlantic routes that passengers could go on elsewhere by train rather than using Heathrow as an airport. The route would be from just outside the Kings Cross or St. Pancras area to Scotland serving Birmingham, Sheffield, Manchester and Leeds as well. The Conservative government re-elected in 2017 has confirmed that it to go ahead. The attraction is it would help to reduce congestion on the other lines as well as being a major attraction in its own right.

One of the problems is how many intermediate stations there should be and where, for example Reading would seem to be a useful interchange since it serves both the West country and also down to Southampton. Clearly if the Channel Tunnel link was extended to Heathrow it might also be sensible to try to link it with Reading.

At the present time rail only has about 10% to 15% share of the market between London to Edinburgh and it is calculated that a high-speed link would reduce the air share by about 75%. It would also help to reduce the imbalance between more people being employed in the South East but with other areas not developing so fast. One of the features however is that it would only be sensible to do this if the railways themselves try to make sure that trains did not get heavier per passenger.

Where Should It Run?

There has been discussion about where such a line should run, if it were to run to St. Pancras or Euston it would link up with the Channel Link. Another possibility might be for it to have such a terminal, but then to go on to Waterloo and possibly Heathrow or alternatively to go to Paddington in the first place and then to go on to Heathrow, but this would duplicate the existing line. If it goes through the Midlands, Birmingham New Street is already overcrowded and there might be scope for it to run to Birmingham International which would link up with the main line.

Thames Link

This line exists at the present time, giving a link between Kings Cross and Farringdon so that passengers can go through from South to North London without having to change. Clearly this is a major advantage to people, particularly those who for whatever reason such as carrying a lot of luggage, being disabled have problems if they have to change trains with the use of escalators or stairs. There are plans to improve the capacity for this. It would help since with the opening of St. Pancras to European passengers it would give access to South London for some of them and also to some parts of North London as well. With the present plans, it will provide a very frequent service.

Crossrail System Now Renamed Elizabeth Line

The then Labour government in October 2007 finally announced that the Crossrail (now renamed the Elizabeth line) system would go ahead, and the £5,000,000,000 project will be completed in 2018. The line goes from Reading and Maidenhead to Canary Wharf in the Docklands and Shenfield in Essex.

There are new Underground stations at Bond Street, Tottenham Court Road, Farringdon, Liverpool Street and Whitechapel. The Elizabeth line trains carry about 1,500 people and would have double the capacity of the Jubilee line during the peak. The funding would come from a mixture of fares, a supplementary increase in business rates as well as from the government. In November 2008 it was announced that BAA was prepared to put in some finance so that it could serve Heathrow.

Javelin Trains

These are new 140 mph trains, which have been on the domestic routes from many parts of Kent to London from 2009. They were also in use to transport people from St. Pancras to Stratford in East London in 7 minutes and were very useful during the Olympic and Paralympics in 2012.

Railways Comeback

The railways after a long period of decline in the UK and most other Western European countries have had a comeback. Currently the railways in the UK have doubled the number of passengers since privatisation in 1994. Much, however, depends upon government controls for example, whether the government will allow enough cash to be spent by Network Rail upon either major electrification schemes or new high-speed lines.

The Eddington Report

The Eddington report by Sir Rob Eddington in November 2006 on the transport infrastructure including railways went against the idea of a new high-speed link between England and Scotland. One of the reports subsequently on the High Speed 2 line ran to over 5000 pages. Many other observers would think that this might be desirable so that not merely would the railways gain but also that the environment would be improved, since it would take away most of the domestic air travel between London and Scotland.

On lines which cannot be electrified the introduction of high-speed trains which have been introduced from the North East of England to South West have been helpful but the trains are now 40 years old and a replacement would seem to be essential. The Channel Tunnel which was opened in 1994 has been successful in eliminating most of the international flights between and Paris and also between London and Brussels.

Contrast with Other Countries and Attitudes Towards High-Speed Trains

The opening of the line to St. Pancras in November 2007 has increased the amount of traffic even further. In other countries such as France with the Train Grand Vitesse system (TGV) has been very successful with separate track for its trains which run at speeds over 200 miles an hour in attracting many new passengers. Similarly, Germany with the InterCity Express (ICE) has also been successful.

This is similar to the Tokaido line in Japan which has a reputation not only for high speeds but also for punctuality. The Advanced Passenger Train project which was electrified was abandoned, but there have been some tilting trains on the British network which is helpful since the original tracks were built to avoid gradients and so in many cases had tight curves. The TGV and many other high-speed trains have almost the opposite specifications (i.e. that gradients are not a problem because of the tremendous power of the trains but they avoid tight curves).

For suburban services, much depends upon the government's attitude towards subsidies. There have been some extensions to the Underground network with the extension of the Jubilee line which has been very successful in obtaining traffic from the car so that very few people in the docklands now travel to work by car. In the London area the Docklands Railway has been extended from the original line which ran from near Fenchurch Street to North Greenwich as to Lewisham and also to Stratford. It also serves the City Airport and is linked to the suburban services at Woolwich Arsenal. This has mainly used existing infrastructure, although because it is driverless it is a very unusual form of transport.

Light Rail Services

In the West Midlands and elsewhere, there have been many more light rail systems and the original one in the Newcastle-upon-Tyne area has been extended. In London the Croydon tramway system from Beckenham Junction to Wimbledon has been successful and some people would like to see it extended but the then Mayor of London Boris Johnson abandoned the plans. How successful the railways are, depends partly upon whether the system of road pricing which was originally introduced in Singapore and more recently in 2003 in the London area is extended to other towns or cities.

Freightliners

Freightliners were originally under British Rail control and then transferred to a joint operation only to be transferred again to British Rail under the 1978 Transport Act. Later they were sold as a part of a privatisation process. They have suffered partly because of the narrow loading gauge of the rail system. The freight through the Channel Tunnel has not been as great as might have been hoped although some people would like to see the Great Central Line which originally went from Marylebone to the East Midlands to link up with the Channel Tunnel route so that freight could be quickly transferred to and from the continent.

Buses

There have been more articulated buses in many towns in the UK and in many other countries these have the advantage of being able to carry more passengers without any additional costs of drivers. If fares can be paid off the vehicle then this reduces the loading time and makes the buses more likely to be viable. The use of the Oyster card and similar systems which means that they can be topped up at local newsagents as well as at stations means that the loading time is minimised.

There have also been more developments in easy access no step buses which means that they are much more accessible for the large number of people who have problems in walking onto the bus.

Improvements to Local Railways

There are areas such as Bristol which still have very little in the way of a local railway system even though in the past there were suggestions for a major tram system. If the metro could share existing railway tracks, as for example with the route to Avonmouth, then this would be helpful. There may also be more scope for the valley lines around Cardiff, although there have been some developments such as a return of passenger services to Ebbw Vale. In the 1980s there had been inner city partnerships which looked at the developments of railways and other modes of transport and their contribution to the inner cities. One of the features of much of the railway system is lack of good information. This is not always helped by operators having many timetables which only show their services, but not other operators. This does not help since in many cases passengers who are not well informed will assume that the local services are less frequent than the reality.

Double-Decker Trains

There has been considerable discussion about this, partly because of the growth of numbers on the railway network. There are however problems if this were to be introduced into the UK. There is therefore a need to compare this, with alternatives such as longer trains. One possibility is that the UK could have longer carriages which are also double deck. This however, would be quite costly since it would need to either lower the track in some areas and possibly to alter tunnels in other areas as well as foot bridges. One of the problems anyway for the railways is that as people get generally bigger that headroom becomes greater unless people are to be uncomfortable. Longer single-decker trains have the advantages that if there are disruptions to the system, that the single-decker trains can be diverted without too much of a problem whereas unless much of the system is devoted to double-decker trains this is not true for them. On the other hand, even longer single-decker trains such as extending them to 16 cars would require considerable amounts of change to stations such as Liverpool Street and Victoria which are two of the busiest stations on the railway system at the present time. A 12-car double-decker train has a similar carrying capacity to a 16-coach single-decker train.

Eurostar

The Eurostar made the fastest trip so far, spending 2 hours from London to Paris in 2007 and also for Brussels. This means that in many cases although this was just a test train that the times between London and places such as Paris and Brussels are already far lower than by air when one accounts for the waiting time at airports. Eurostar is currently developing from London to Amsterdam.

It would be possible to link this in with other developments such as that from Paris to Cologne which would also mean that even allowing for a transfer that this would be quicker than by air and Deutscher Bahn wants to develop this.

Government Policies on the Railways

In July 2007, the then Labour government announced that the plan for the railways was that fare holders should pay more so that the subsidy would be reduced from £4.5 billion a year to £3 billion a year. Whereas at the time of privatisation the assumption was that passenger numbers would go down, the government estimated that by 2014 the passengers would make 180 million more journeys a year, compared with a total of around 1.1 billion.

The government generally set a higher target for punctuality so that the aim was that the bench mark should be 92.6% compared with the current 88%. Two potential bottle-necks (i.e. at Birmingham New Street and at Reading were given government money with £425 million being given to Reading and £120 million towards Birmingham New Street). There was also an announcement that there would be price caps on season tickets and saver fares and that there would be 13 hundred new carriages by 2014. On the other hand, no decision was to be taken on a new North to South railway before 2012.

Whilst the UK government has announced that diesel and petrol cars will be eliminated by 2040 has been no announcement about eliminating diesel lorries. Whilst most people notice what is happening with exhaust fumes from lorries, 90% of some pollution occurs in other ways such as breaking down of tyres and brake linings. This is particularly important to Kent Southeast England where both the channel tunnel and the ferry ports generate a large amount of traffic. Pressure groups such as Rail Future have recommended that there should be an annual conference between Eurotunnel, local authorities in Kent as well as the DfT to examine possibilities of transferring freight from road to rail.

Franchises

Many people have suggested that the current length of Franchises of about 7 years is not helpful since it does not give the operators time to go in for radical solutions such as thinking of electrification or even of the best use of carriages and locomotives since much depends on the ROSCO.

Many people would also suggest that there should be more use of section 106 agreements, which means that if there is a planning gain that some of the money could be put back into rail schemes rather than other forms of public expenditure. There is also a need for greater integration between cycling and railways. This could be by better use of cycling lockers at stations.

Light Rail

There has been more interest in light rail and tramway systems and by 2007 around 90 new systems had come to being. One of these has been in the West Midlands, where the Merry Hill shopping development owners have offered £35 million, since it is helpful to the running of the new shopping centre. However, there have often been criticisms of the procedures since for example whereas there were to have been a South Hampshire rapid transit system this has not happened, similarly there has been problems in Manchester as well as Leeds. Part of the problem has been the increase in cost, since the idea was suggested, and also part of the problem has been that governments have sometimes changed their minds about what should happen. On the other hand, the advantages of trams are that they do reduce the number of people coming in by car, and the estimate is that 20% of peak hour travellers previously used cars and 50% of weekend travellers do so. Economists can see some of the effects in London where traffic growth in Croydon was reduced quite drastically whereas in nearby Kingston the numbers fell much less significantly. It has also resulted in higher employment rates in the more deprived areas which the Croydon tram link has served. Whilst shopkeepers have sometimes objected since they think that it will not be helpful to them, Croydon Tramlink has shown that the average footfall has risen by 11% and the average spend per customer has risen by 12%.

Improvements to Railway Stations

It is difficult to find out the exact number of disabled people in the UK. It partly depends upon what we mean by disabled. The cost of new stations on the Jubilee line which has very good access for disabled people has been extremely high, but of course the disabled in many cases have very little access to existing Underground stations or to many railway stations. The use of lifts or escalators or even if this could not be afforded the equivalent of a small conveyor belt alongside the stairs at stations would mean that people could carry luggage much more easily. This would be helpful as well to other groups such as women with small children and shopping.

Stations to Be Less Separate from Rest of Town

One of the problems with many town centre developments is that the railway stations have been separated from the central high street areas by main roads. This is one of the many reasons why people may be reluctant to use the train rather than the car. It would seem sensible in the future if the stations can be better linked to the centres. It is noticeable in Chester that there is a free bus service from Chester station to the centre of the city and this is financed by a number of different railway companies amongst others.

Rural Stations

There have been some improvements in rural areas carried out by so-called community railways. There are over 50 community rail partnerships and there is also an Association of Community Rail Partnerships. One suggestion might be that the stations would not just issue tickets, but might be able to sell other items particularly at times when trains where not there.

Community Railways

These apply where the railways have limited amounts of traffic. The idea is that local authorities, organisations and individuals come together to form community rail partnerships. The DfT defines a community railway as one where there are not any major freight flows, and where one main operator provides the bulk of services and they are not serving a major conurbation directly. Sometimes they have been very successful for example on the Saint Erth to Saint Ives line a regular shuttle service increased demand by 50%.

Road Transport

In 1980, the Armitage committee looked at the problems of heavy lorries. It had the largest number of submissions, over 1,600 for a transport enquiry.

The report eventually led to the introduction of the 38 ton lorry. The report itself had suggested a 40 ton lorry, although with safeguards such as the use of improved construction and use regulations such as rear guards and side guards.

There were little disagreements about the possible economies of scale for the road haulage operators including a reduction in fuel per tonne kilometre and also in driving costs since one driver is needed whatever the size of the load. There was, however, much more dispute about whether it would lead to an improvement in either safety or environmental aspects.

Within conurbations the Wood enquiry tried to set out the costs and benefits of banning lorries at different times of day, although it was not allowed to look at road pricing which most economists would have thought was desirable.

In 2008, there were proposals by the road freight lobby to run lorries weighing up to 60 to 84 tonnes and which would have an overall total length nearly 9 metres longer than the current maximum permitted length.

The Rail Freight Group (RFG) wrote to the Ministry responsible saying how unpopular the proposals would be and also how much more pollution would arise as a result since tonnes of freight carried by rail produces at least 80% less carbon dioxide than by road. It also pointed out that the average freight train removes 50 lorries from the road network.

There might be scope which the DfT now tries to publicise through better use of computer scheduling and also of operational research techniques which would reduce the number of lorries in any case. Some firms such as Tate & Lyle have tried to improve the utilisation of their vehicles by obtaining back loads. Transport Statistics show that many vehicles still travel empty on the return journey and clearly this is not desirable for either financial or environmental reasons. At one stage, the government had an automatic increase in fuel tax but this has been partly abandoned because of protests as in 2001. The government tried to increase the number of battery operated vehicles by eliminating the vehicle excise duty in 1980. Many people feel that whilst the electric vehicles have always been used for milk rounds there could be increased scope where vehicles do little mileage during the day. One local authority, Westminster has tried to encourage electric vehicles by allowing free refuelling in their area. Electric vehicles whilst more expensive initially have a much longer vehicle life which might well offset this cost. Other car manufacturers have experimented with hybrid vehicles which can run on either battery or conventional fuel. The road pricing system in London in 2003 might encourage more people to consider alternative fuels, since then they do not pay the congestion charge.

Shipping

In late 2007 just before the credit crunch more trade with both China and India meant that there was much greater demand for shipping and also a shortage of some types of ships so that some ship owners have converted oil tankers into dry bulk carriers. Part of the reason is that globalisation means that goods get sent much further than in the past. China itself is aiming to be become the major shipbuilding country in the world replacing South Korea and Japan in the top slots.

One of the features of the shipping market, has been that the volume of containers sent by ship has been growing at a very fast pace. This has been partly because there have been fewer trade barriers as a result of the World Trade Organisation efforts. It has partly been that there has been a transfer of manufacturing to cheaper bases which has meant that manufacturing is often carried out in relatively low-income countries notably in Asia, but then the products are sent longer distances to the higher income countries such as Western Europe and the USA. It has also been partly because there has been a general rise in income.

One of the other features is that there have been an increasingly a small proportion of ports which cater for this traffic.

The port capacity needs to be addressed on a world-wide basis, since clearly over capacity will lead to waste whilst too little capacity will mean that there are major hold-ups at ports.

Part of the problem here is that both the assets at the ports and the ships themselves have a long life, and it is difficult to forecast trade with accuracy.

Another feature has been that the conference system has come to an end and this has a number of different effects. One of them is that it makes life more unstable in many ways and also that shipping companies which have been used to giving information to each other no longer do so.

In order to forecast the future, one would need to look at the major ports, showing what capacity, they have and where the traffic goes to and from. It is also possible to look at the type of trade which is carried through these ports and it seems likely that the tendency will be to look at economies of scales of ports.

One of the other features in the shipping market has been that a large portion of it, just over half is in chartering currently. Chartering means that the ships are in effect leased for a period to the people who wish to send the goods.

The strong tendency has been for the growth to be concentrated on the Far East routes and in particular we can see that certain commodities such as telecoms and road vehicles have been amongst the rapid growth markets. The rapid growth in container traffic is not solely due to an increase in the growth of the goods market, but also because people have switched from other methods to containerisation and also more empty containers is sent on the return journeys.

The increased cost of fuel may have the effect on a greater emphasis on slower saving speeds, which reduces the amount of fuel used. It may also mean that because the ships will not go so fast that there could be an increase in demand for shipping capacity.

Economists already know the current capacity and given that ships take a while to go from the ordering stage to launch they have at any one time a rough idea of the capacity for a short period.

The ports themselves try to compete in a number of different ways. One of them as we have seen earlier is that the locations of the ports themselves have changed so that they can offer greater draughts. The ports themselves need to have however good connections with rail and road and it is noticeable in the UK that about 25% of the container traffic is then transported on by rail. Ports themselves may often tranship containers to other ports.

The 64 major ports currently handle just under 70% of the global throughput.

It seems likely that even if the major ports were all operating with maximum efficiency that there would still be a need for expansion of the major ports. However, some care needs to be taken with this assumption since a greater emphasis on the environment could mean changes in the fuels which are used in the future and of course fuel itself is one of the major components of transport demand although not by container ship.

Whereas in the past the British Government owned indirectly many ports some of which were taken over by Associated British Ports and the government also operated a Dock Labour Scheme the government now has little direct effect on the ports although clearly the road and rail system have been altered in some cases to meet the needs of ports.

When looking at ports capacity the ports themselves need to look at the environment and it is noticeable that the Debden terminal (near Southampton) failed to get consent. The government itself published a report in July 2007 which suggested that overall port traffic would only grow by 1% per annum to 2030, but the roll-on/roll-off traffic (RORO) would increase considerably as would deep-sea container traffic.

The likelihood is that about half of all rail freight traffic is related to imports and exports so that ports policy cannot be decided in isolation from the rest of transport policy. It also seems likely that demand for coal will fall, although in October 2007 it was announced that there will be a new coal powered power station in Kent.

Air Transport

The A320 is the second best-selling aircraft of all time after the Boeing 737. At the time of writing, it had sold over 8,600 Airbus A320 family aircraft. It was the first commercial aircraft to use fly by wire control systems. The A380 is a double-decker plane and can hold over 800 people if they all use tourist class or over 550 people if they have a standard 3 class configuration. It is the largest commercial aircraft and is intended to take on the might of the Boeing empire which as can be seen from almost any visit to an airport tend to have dominated the worlds' airlines. It has had many teething problems and it remains to be seen whether it is a vision of the future.

Taxation

At the present time, passengers pay a flat rate tax in the UK for air travel. This seems totally illogical. If the tax was imposed upon the capacity of the aircraft, it would seem more sensible since then the operators would have more incentive to have a better load factor. At the present time, an empty aircraft would have no taxation imposed upon it whereas a full one would. This was altered from 2009.

Expansion of Heathrow

In November 2007, the government announced that it favoured an expansion of Heathrow with another runway although most of the local authorities in the area were opposed to this. The government announced plans for a third runway in 2009. In 2016 the government stated that it favoured the expansion of Heathrow rather than Gatwick.

Open Skies Agreement

This was signed between the EU and the US in 2007. One of the main airlines involved in such flights is BA who have a large number of slots at Heathrow for historical reasons.

There has been an increase in the number of airlines, the most notable of which is probably Ryanair. Ryanair has kept to its no frills policies whilst some of the other low-cost airlines have competed on a more straightforward basis directly with the conventional airlines.

The vacant UK slots are likely to be taken up by American airlines. Much, however, will depend upon the price of the dollar. Clearly in 2007 when the price of the dollar fell so that there were over $2 to £1, it was much easier for the US airlines to compete and they thus might wish to expand into European markets including the British one.

It seems less likely, however, that many European lines including the British would want to expand into the US market where profit margins are low. There might be some exceptions as with the business only aircraft where they could possibly gain a reasonable market if they could go out at early hours in the morning to catch the previous market which was held at the stage when Concorde was flying.

One possibility would be for EU airlines to take over US airlines. If economists look at the current operations between the EU and the USA, the vast majority of routes are between the East Coast of the USA such as New York, Miami and the EU with very few going as far as the West Coast and there seems little reason to believe that this will change with open Skies.

More globally the Indian market is fast growing with the rapid increase in incomes on the huge Indian market but as the Indian travel market is one of low profit margins it is not clear that many EU firms will wish to expand into this market if they do not have a presence at the present time.

Regional Transport Strategy for South East England

In the South East journey lengths by car are decreasing and the regional transport strategy (RTS) suggests that this is partly because in the peak hours, there are limits on capacity and also that people can often find facilities nearer to home as a result of land use planning. It also suggests that more people either work from home or use remote working. It does not however mention other ideas such as that more people may be able to buy some items on line including groceries and clothes which means that road freight is substituted for the car. The RTS does make the point that the region does not have to be passive but can alter some of the factors of demand.

Rural Areas in the South East

Whilst the South East has the highest rate of car ownership and the car is predominant there may be ways of trying to reduce demand by car and also to give choice to people who have little or no access to cars. This might be done through improving provision for cyclists and pedestrians and also through use of community based transport. It might be argued, however, that the government through the post-office closure of many rural post-offices in Kent and elsewhere is not helping to provide this balance.

Regional Spokes

The RTS wants to see better integration of rail and express bus/coach services along corridors and also to try to eliminate bottle-necks. Whilst the report does not say this the railways clearly have some bottle-necks such as those through London Bridge and also at Welwyn Garden City. Since that period London Bridge has been extensively remodelled and similarly Victoria and Waterloo as well as St. Pancras, Kings Cross and Paddington in the London area are being transformed.

Road Pricing

This states that local authorities might consider new charging initiatives if they will help to deliver a better transport policy. Currently some authorities are considering different parking charges for different types of road vehicles.

Mobility Management

It suggests that there should be more use of incentives for car sharing. Although the report does not state this, one way to do this might be to reduce the number of car parking spaces, which either firms or local authorities have.

It also suggests that there may be better measures to increase accessibility to railway stations. One method of doing this, would be to ensure that railway stations are considered when developing town centres, rather than in many cases putting in effect a bypass around the high streets which often severs the link between the railway station and the high street.

Parking

Planning policy guidance note 13 (PPG13) requires development plans to have maximum levels of car parking for different types of development. The RTS suggests that local authorities should impose lower limits than this, although it would wish to increase railway parking, especially if the railways are part of a regional hub. It does, however, wish to see that access to stations should be improved by other methods of transport.

Ports

It suggests that there is more scope for development of Southampton. It also suggests that there is a greater need for improvement on roll-on/roll-off ferries at a number of ports, such as Ramsgate and Southampton. It does not seem to be in line with other commentators that there should be much greater emphasis on the Channel Tunnel rather than building up the traffic at the Channel Ports. It also suggests that there is a greater need for deep-sea containers such as Southampton and Thamesport.

Airports

RTS had strong reservations about the 2003 White Paper on aviation and suggests that more traffic around Heathrow would have a high impact on pollution as well as on trying to accommodate traffic to and from Heathrow. It suggests that this will be considerably true of a third runway and Terminal 6 at Heathrow. It suggests that there might be more scope for regional airports such as Southampton and Manston.

UK Government Review Backs London Heathrow and London Stansted Airport Strategy

The government suggests that there is a need for more runways at both Heathrow and Stansted. This could be done by having a second runway at

Stansted to be built between 2011/2012 and also a third short runway between 2015/2020. It does however make a point that it would need to take account of meeting air and noise limits. We might note here that the government thinks that more runways are necessary.

Alternatives to Air Transport

Many commentators might feel that this is not the way forwards and that the need for example of short term and longer-term circumstances are not compatible with the idea of global warming. whilst clearly in the short term, the USA run is not really suitable for other forms of transport but if as hoped by the existing Russian government that there could be a Bering Strait tunnel, this would open up the idea of a long-distance rail route which might be attractive for some passenger travel and perhaps more important would also open the idea of freight from Europe to the USA in timings which whilst slow could well be ideal for many west coast US routes.

For short distance air routes such as those from the UK to France, Germany and the low countries the prospects of even more TGV trains in France plus even more fast routes in Germany with ICE trains or the equivalent will mean that for many journeys the overall times difference between trains and air is relatively small. Therefore, much will depend upon the pricing policy of the railways and the pricing policy of the airlines which in turn will depend partly upon the taxation policy of the governments concerned including the UK government.

October 2007 Pre-Budget Report

The government announced in this report that in the future the duty would not be on individual passengers but on flights, which would encourage airlines to make better use of planes. Clearly at the present time the duty of £10 per economy class ticket is likely to have an effect on low cost airlines, but not upon more expensive airlines. It might have the effect of encouraging higher price airlines to look for better load factors and the use of the internet might mean that lower fares might be available the closer it was to the time of the departure for less full airlines. It was also expected that the tax component for example of going across the Atlantic might increase to between £40 and £80, whilst for Europe business and first-class passengers would be charged £20.

Rail Freight

It suggests that rail freight should be improved, possibly by looking at bottle-necks and also suggests that there are a number of places where bottle-necks exist which would if eliminated help to increase rail freight.

Greater Use of Railways

One of the problems is that whilst the EU is often blamed for little spending in the UK, in many cases EU money has been spent on the West Coast main line as well as for freight at Felixstowe and also on the Channel Tunnel rail link. It might be possible to have a new line linking Canvey Island in Essex and North Kent so that there could in effect be a freight bypass for London.

Rail Franchises

Many people have suggested that the current length of Franchises of about 7 years is not helpful, since it does not give the operator's time to go in for radical solution such as thinking of electrification or even of the best use of carriages and locomotives since much depends on the rolling stock operating company (ROSCO).

Many people would also suggest that there should be more use of section 106 agreements, which means that if there is a planning gain that some of the money could be put back into rail schemes rather than other forms of public expenditure. There is also a need for greater integration between cycling and railways. This could be by better use of cycling lockers at stations.

Oyster Cards

Oyster cards have become very common in the London area, one of the features is that because it automatically adds up the sums spent in a 24-hour period between 4:30am and 4:30am the next day. If people are not quite sure what is happening they can buy single fares and the amount they pay will always be less than the price of a one-day bus pass or travel pass.

Glossary of Transport Economics

Accelerator – The accelerator describes how the capital goods market fluctuates with changes in demand. Given certain assumptions, it can be seen that a relatively small change in demand for final products can lead to a much larger change in demand for capital goods. This is relevant to the transport industry especially where there are long lasting assets. The shipbuilding industry is very subject to these fluctuations. This can mean that if there is increasing demand for shipping, the shipping industry may find that there are difficulties in trying to obtain the ships that can be brought.

Allocated Costs – These are costs allocated to particular services. Because there are often joint costs particularly on the railways the way in which costs are allocated can be arbitrary, for example in British Rail days then if inter-city services were regarded as a prime user then they were charged all of the related costs. There are a number of alternatives, they could be allocated on a train basis or per unit carried.

Average Costs – Average costs can be defined as:

$$\frac{\text{Total Costs}}{\text{Total Units}}$$

Average costs are sometimes used as a method of comparison between different operators to judge efficiency. Where the average cost curve is at its lowest point is often referred to as economic efficiency. This is quite different from technical efficiency. The average cost will be at its lowest point in perfect competition although not generally in an oligopoly or monopoly.

Average Cost Pricing – This was widely used on the railways after the Transport Act 1947. This was done in order to avoid fears of discrimination but tended in practice to lead to cross subsidisation, since rural routes had higher costs per passenger mile or tonne mile than urban routes.

Average cost pricing was also common on most road passenger undertakings in the UK prior to the increase in competition following the 1980 and 1985 Transport Acts.

Avoidable Costs – This is a term, which was used in the Railways Act 1974. Avoidable costs are those costs, which would not occur if the service were run. Therefore, it is approximately the same as the economist's concept of marginal costs.

Balance of Trade – This refers to visible trading for both imports and exports.

Balance of Payments – This refers to both invisible and visible trading for both imports and exports.

Barriers to Entry – This term has often been used in connection with monopolies and oligopolies. The barriers to entry in the transport industry may be licensing systems such as that which existed in the road freight industry in the UK prior to 1968 and which still exists in many other countries. The barriers to entry may be the high cost of capital equipment as in the oil tanker industries.

Betterment – This refers to any increase in the value of an asset, which could include land. This could occur with land being used for airports where it is originally drained to make sure it is suitable for its purpose.

Breakeven Policy – This may be a business objective for the public sector and was laid down in some transport acts noticeably the Transport Act 1947. It was also an objective of many municipal bus undertakings on the assumption that it was felt unreasonable that bus undertakings should make a profit from passengers who would usually be the then local rate payers but neither should they be subsidised from the rates. This policy was quite often altered in the 1970s partly because of increasing costs and was altered again by the 1980 and 1985 Transport Acts.

Breakeven Volume – This can be found in principle by working out the fixed and variable costs to get to total cost and then to work out the likely revenue for each unit. From this one can then find out the breakeven points. However, in most cases, this is not very helpful since we can often sell more if we reduce the price. Where, however, charges are laid down such as for taxis by many local authorities it could be of some use.

Business Objectives – Whilst many economists have assumed that profit maximisation is the major objective, it is by no means the only objective of all transport firms. The public sector objectives may be more complex than this, whilst in the private sector growth maximisation or satisficing may be the objective.

Capacity – The capacity of the system is the maximum amount of traffic, which the system could carry if all the factors of production were being fully utilised. The capacity of the system is usually expressed in terms of a number of passengers, passenger kilometres, tonne kilometres or tonnes that can be carried per unit of time. Often the total capacity of a system is only being used in the peak times. The capacity of the system varies according to the size and speed of vehicles as well as loading and unloading times. The capacity of the system may be outside the operator's control as with road passenger transport whereby a system of bus roads or bus lanes might improve capacity or may be within the control of the operator as with the former British Rail. In recent years there has been an increasing tendency to try to improve freight capacity by speeding up loading and unloading.

Capital – This is a term usually used by economists to denote goods which are used to provide other goods or services. Thus, buses to a bus operator will be an example of capital goods but to a bus manufacturer would not be.

Capital Gearing – Companies are said to be highly geared when they have a high percentage of fixed interest capital to equity capital. Private sector firms with high capital gearing are more vulnerable to take-overs than other firms.

Cartel – This is the name given where firms combine in such a way that the total output and pricing of the cartel is provided for collectively rather than through the market. For cartels to be successful there must usually be barriers to entry for other firms and demand should check for the mode of transport as a whole should be inelastic otherwise there will be little point in restricting output. Organisations such as International Air Transport Association (IATA) and shipping conferences whilst sometimes thought of as being cartels will both deny they are.

Cash Flow – This is the net amount of cash flowing into and out of the business during a specified period. It is important in transport economics especially where the amount of finance is large and long term assets are being brought. This is because an undertaking could be profitable in the long term, but if there is insufficient liquidity in the short term the undertaking could still go bankrupt.

The Channel Tunnel – One of the features of the Euro tunnel is that it operates its own shuttle trains which carry cars and coaches as well as freight on a figure of eight loop between the UK terminal at Cheriton and the French terminal at Frethun. These shuttle trains have a larger loading gauge than either the British Railway operators or the Société nationale des chemins de fer français (SNCF, "French National Railway Company") and therefore they would not operate on either the British or the French railway systems.

The other half of the capacity of the tunnel was leased to the then British Rail and SNCF and this operated a system of high-speed passenger trains.

Clearing House – This is the name given to any organisation particularly in road freight where information is brought together regarding both available traffic and also vehicles. The clearinghouse would usually receive a commission. Prior to nationalisation of the railways, there was a railway-clearing house, which had a similar function.

Collective Goods – This is sometimes also referred to as public goods. This is where more people using the facility do not use up goods or services. The opposite of collective goods is a private good such as food where one person eating food means it is not available for another. In the transport system traffic lights, street lighting and lighthouses are examples. Such services have not generally been provided by the private sector since it is difficult to devise a price mechanism, which does not waste resources. The marginal cost almost by definition of a collective good is zero.

Competition and Markets Authority Originally the Competition Commission – The industry as a whole was governed by the then Competition Commission. In May 2007 it organised an enquiry into how we could model competition in the first place. It was also looking at the rolling stock operating companies (ROSCOs).

Computers and Transport – As with other industries computers have been widely used for accounting purposes. The vast number of calculations that computers can do very quickly makes them extremely helpful in the issuing and checking of invoices to prepare dispatch notes, etc. They could not only be used for bookkeeping purposes but also are increasingly used for costing purposes. In the road freight industry, this has been made easier with the data which is now available from tachographs. It is therefore much easier than before to allocate repair maintenance and fuel costs on the sensible basis. Computers are also widely used for stock purposes. Given the lead time for deliveries it becomes easier to hold the right amount of stock. Computers have also been used in transport as well for payrolls. This has considerable advantages especially where overtime and bonus payments have to be made which is fairly common in the transport industry.

Whilst computers have generally until recently been used in effect as a large calculator the more sophisticated use of computers is now taking place. This is the ability of computers to be able to transfer data quickly from one purpose to another. Computer programmes for example in the road freight industry have been developed which will show the distances that have been travelled, the timing of journeys, the amount of down time (i.e. the time which the vehicles are not being used for and why this has taken place). This should enable the road haulier to more effectively schedule his or her vehicles. This should also enable the haulier to look at the overall life cost of the vehicle and thus to be able to make more effective investment. The ability to see from tachograph data using computers gives clues as to reasons why productivity has not been as high as it should be, for example it can show whether delays were caused at the depot or at the shops to which the vehicles are going. This data should enable managers to manage much more effectively.

Complementary Goods or Services – These are goods or services where the demand for one rises or falls in line with demand for another good or service. An example in the transport industry would be fuel and motorcars or that of a railway branch line, which acted as a feeder to an Inter-city service. Complementary services could go across modes (e.g. the Heathrow Express is one for many flights from Heathrow).

Concentration Ratio – This is used to define the extent of monopoly power. The Competition Commission would usually only investigate monopoly if a firm has 25% or more of a market. Concentration ratio is often used to investigate the share of the market by say the largest 3 or 4 main firms.

Consumer Surplus – This is the difference between what a person is willing to pay and how much he or she does pay. Clearly if people get free fares as UK people over 60 do on off-peak buses in their locality they would almost certainly have been willing to pay some money for the journey beforehand. Therefore, if a government wishes to know how much benefit they are getting it will be important for government to have some idea of what are the ensuing benefits.

Contestability – This is the ease with which new firms could either enter or leave the market. How easy it is to enter the market could depend upon private firms not engaging in predatory pricing. It could also be prevented by governments which in some cases have prevented new firms entering the market sometimes through a system of quantity rather than quality licensing. In other cases, it could be the indivisibility of investment which prevents new firms entering. For example, even if the Eurotunnel did not have a long term franchise from the British and French government, it is difficult to see that there would be many other firms wishing to enter the tunnel or bridge market from the same destinations.

Cooper Brothers' Formula – This was a formula that had been used by Cooper Brothers to show what grants should be available for rail services which the government thought should be provided for social reasons, under the Transport Act 1968. It was not necessarily realistic since it assumed that some elements of joint cost could be saved if the service was withdrawn. This was found not to be true and therefore the concept of avoidable cost came in with the Railways Act 1974.

Correlation – This is a statistical method of measuring the relationship between one set of variables and another. There can either be ranked correlation (i.e. where economists measure variables in orders of preference or a product moment correlation which is the more common one in economics). The fact that there may be a strong relationship between one item and another does not prove causality. Many people have assumed that people buying more cars is a method of increasing and improving prosperity whereas some past reports from the World Bank would dispute this. A positive correlation shows that one variable goes up in line with another (e.g. generally incomes and demand for air travel whilst a negative correlation is where one variable e.g. demand is inversely related to another e.g. price).

Cross Price Elasticity of Demand – This may be defined as:

$$\frac{\text{The percentage change in demand for X}}{\text{The percentage change in price for Y}}$$

It is positive for substitutes for example to bus services and negative for complementary goods or services for example petrol and cars.

Cross Subsidisation – This occurs when profitable services make up for losses on unprofitable ones. This occurred widely in bus services particularly in municipal undertakings. The cross subsidy here was sometimes between urban and rural routes but sometimes between the peak and the off peak.

Deficit Financing – This occurs when transport undertakings have run at a loss and then had the deficits made up by the government at the end of the period. It has in the UK been replaced by the Public Service Obligation.

Demand Curve – This is one of the most important concepts in economics showing how much the demand is at any given price.

Demonstration Effect – This is usually associated with a new service or product and to the first stages of the product life cycle. It means that once people have seen something being used (i.e. demonstrated that they are more likely to go out and buy the goods or services). Apart from the direct effect on transport there can be indirect effects (e.g. if people see that a computer program is useful in the transport industry then their competitors may well see this and go out and buy it).

Depreciation – Depreciation is one of the costs of business and is particularly important in the road haulage sector. Its assets will lose value over time partly because of fair wear and tear. It is therefore important to set aside money for replacement at the end of the commercial life of the asset. The commercial life of the asset may be less than the technical life. For example, on British Railways steam engines were still being built until 1960 with a technical life which might have been 40 to 50 years. However, they were all replaced by 1968. Here the commercial life is therefore very much less than the technical life. There are several different methods of calculating depreciation. One of them is the so-called straight-line method. If for example a road haulage vehicle was bought for £11,000 and had a life of five years and a resale of £1,000 at the end of the year. The depreciation would be £2,000 per year (i.e. the amount of the annual depreciation equals):

$$\frac{\text{Original cost minus residual value}}{\text{Lifetime of asset}}$$

The second method of depreciation is by the diminishing balance method. In practice, motor vehicles particularly company cars will lose a great deal more retail value in the first year than subsequently. Also vehicles including motor vehicles will require relatively a small number of repairs in the first few years but after that breakdowns will become more common. By having a diminishing balance, for example 20% per year there can be equal instalments of money charged for depreciation and larger repair bills.

If we take for example a charge of 25% on diminishing balance and the asset was worth £34,000 at the start, then we would depreciate the item by £8,500 in the first year leaving a balance of £25,500. In the next year we would depreciate the balance of £25,000 by 25% (i.e. £6,375). We would then continue with similar calculations for the third and fourth years.

An alternative method of depreciation is by the revaluation method. Some assets do not necessarily depreciate steadily over a lifetime. In this case they are valued at the end of the year, if the value is found to have increased over the year as for example with cost of buildings the extra value can be taken as a profit whilst if it has fallen the decrease is shown as a loss.

The fuller discussion of these points is outside the scope of an economics book but can be found in any introduction to accounts including the authors' book "Principles of accounting".

Derived Demand – Transport is rarely demanded for its own sake but usually because it gives utility of place and time. Therefore, it follows that demand for transport is often linked with changes in location whether of housing by passengers or by those of firms producing goods and services. There has in recent years been more interest in the location of goods services and also the relationship between land use planning and public transport.

Diminishing Marginal Returns – This concept is the idea that if there is a fixed factor of production adding other factors production may give increasing returns at first but will eventually give lower returns. The common-sense appeal behind this is that if this was not true we could gradually build a bigger and better warehouse, depot, etc., clearly this is not the case. In the case of the railways, for example if we have single track and double it we may more than double the capacity and if demand is sufficient this may well be worthwhile. Adding a third track, however, may not be worthwhile.

Diminishing Marginal Utility – This is the basic concept behind the demand curve. It refers to the idea that utility will decrease with more units consumed. In transport, the ideas will almost certainly hold. Travellers may gain considerable utility from some journeys, such as to and from business but derive much less utility from other journeys. Even people whose marginal financial cost is zero as with season tickets will not wish to spend all of their time travelling.

Direct Costs – This is a term often used by accountants and is very similar to the economist's concept of variable cost. These are the costs which occur directly because of a particular activity, for example the direct costs of an additional aircraft journey would be the fuel and oil overtime payments to crew members, etc.

Discounted Cash Flow – This is a method of investment appraisal. The logic behind the technique is to try to value all future benefits and costs in some common denominator.

Diversification – This is where a transport firm moves into activities which are new to the organisation. It may be linked to horizontal or vertical integration. In these cases, we can see that security of supply or for outlets may be useful to the organisation, but not necessarily to the customers. In other cases, the firm may be a conglomerate i.e. having interests in unrelated activities (e.g. Virgin with its interests in rail and air but also with entertainment is an example).

Economic Rent – This is the factor earnings which are over and above those which are necessary to keep a factor of production in its current use. It will usually only apply if there is a degree of monopoly or monopsony power.

Economies of Scale – This refers to the advantages of being larger. It could occur in terms of a firm, a depot, a network. The basic concept behind the economies of scale is that of specialisation of either equipment or labour.

Economies of Scope – This is where a firm enlarges itself by acquiring other firms in the same line of business.

Elasticities of Demand – This shows how demand may alter because of changes in different variables. We can have price elasticity of demand, income elasticity of demand or cross elasticity of demand.

Elasticities of Supply – This shows how supply may alter because of changes in different variables such as price, if an organisation is near to capacity as with the peak hours in most towns there may be very little elasticity in the short term but rather more in the longer term.

It may be defined as:

$$\frac{\text{The percentage change in supply of X}}{\text{The percentage change in price of X}}$$

Entrepreneurs – These are people who take the risk for the business. The term came into prominence in the nineteenth century when the entrepreneur was often a single person or family who put in the money and also controlled the organisation which they had set up. In many large organisations nowadays, it is difficult to identify the entrepreneurs.

Equilibrium – This is the position often discussed by economists whereby the demands of consumers are equal to those of suppliers. It is usually assumed that the interaction of the users and the suppliers will result in this state. It may take a long time to come about in transport if it ever does partly because much of the transport infrastructure as well as many of the vehicles has long lives so that the time periods of operation may well be considerable.

EU (Sometimes Formerly EEC or Common Market) – EU stands for European Union whilst The EEC stands for European Economic Community now more commonly referred to as the European Community or European Union. In former times it was more frequently called the Common Market. It had fifteen members up to 2004 (i.e. France, Germany, Italy, Netherlands, Belgium, Luxembourg, Eire, Denmark, the UK, Spain, Portugal, Greece, Sweden, Austria and Finland). A further 10 entered in 2004 Czech Republic, Hungary, Poland, Slovakia and Slovenia from the former Soviet countries as well as the 3 Baltic states of Estonia, Latvia and Lithuania and the two former British colonies of Cyprus and Malta. It grew to 27 as Bulgaria and Rumania entered in 2007. The EU has a common transport policy, which has affected road freight, with its tachograph regulations and hours regulations, both of which were modified in 1985.

Excess Capacity – This is common in the transport industry especially during the downturn in the economy which often happens as part of the trade cycle. It can also happen when restrictions are imposed upon the transport industry. Embargoes can be imposed for political reasons such as sanctions in 2017 against Russia. This will lead to excess capacity on transport operators to and from Russia.

External Financing Limits – This was the maximum limit, which was laid down by the government about how much nationalised boards could have outstanding from outside sources in any one year.

Externalities – A great deal of controversy in transport occurs because of externalities, which are costs or benefits, which are not paid for directly by the consumer of the goods or services. Externalities can be favourable for example householders having the value of their land increased as a result of transport improvements or unfavourable for example increasing noise, pollution, congestion or accidents from a badly located road or airport.

Extrapolation – This is one of the most common forms of forecasting demand. It usually assumes that with a regression model that we can continue to make the same assumptions for the future. This is not necessarily helpful for long term forecasting where shocks to the system can radically alter demand or costs.

Factors of Production – Economists have often referred to three factors of production (i.e. land, labour and capital). Land does not just mean land in the conventional sense but also includes the seas, rivers and oceans which is obviously important for the shipping industry. It also includes minerals under the earth. Capital is a good used for the production of other goods and services. Many economists have also used a fourth factor of production (i.e. entrepreneurs).

Fiscal Policy – This name is used for government policies on both taxation and subsidies. Subsidies to transport are very common in most developed countries. Taxation is equally important and may alter the modal split as well as influencing total demand.

Fixed Costs – These are costs, which do not vary with output. For the motorist an example is the vehicle excise duty.

Forked Tariff System – This is the system formerly proposed by the EEC whereby minimum and maximum rates will be set for road haulage on international journeys.

Generated Traffic – This is traffic, which is increased as a result of transport improvements. The electrification of the East Coast main line in the 1980s meant that there were far more passengers travelling to and from Peterborough every day. One of the problems in transport economics is that improvements particularly those which try to reduce the urban peak problem may generate sufficient traffic to outweigh the advantages of low fares or better roads.

Goodwill – This is an accounting term and is used where a firm taking over another organisation pays more than is strictly necessary for the assets in order to acquire the goodwill which the firm being taken over has acquired and where the name of the organisation is in itself helpful in order to gain more business.

Historic Cost – This is the cost originally paid for an asset. The difference between this and the current value is measured by depreciation. In times of inflation some assets are likely to go up in terms of money value if not in real terms. This is particularly true of land values. This is one of the main reasons why if historic costs are stated in the accounts there may be differences between this and the current price which asset stripping firms may wish to take advantage of.

Holding Company – This term has been used in both the public and private sectors. The Transport Holding Company existed in the public sector. Holding companies often have interests in more than one country and in more than one mode of transport. This would be true of Arriva.

Horizontal Integration – This is where firms producing similar services either merge or are taken over to give potential economies of scale. British Airways merged with Iberia in 2011.

IATA – The International Air Transport Association (IATA) was formed in 1945 and the majority of scheduled flight operators belong to it. For a long while it played an important role by having an agreed system of fares for scheduled flights.

Indivisibilities – Investment is often indivisible and therefore poses problems to the investor. If Network Rail wishes to do major upgrading of its track it could do this by expanding the number of tracks in some areas. There is not one easy method of recovering these prices from the different rail operators in a manner which is reasonably fair. The same is even truer of the Channel Tunnel.

Inflation – This is a term to mean a general rise in price rather than the usual fluctuations in prices as a result of changes in either demand or supply. It can be measured in a variety of different ways but most typically these include looking at a basket of goods and service over a time series. One measure in the UK is the retail price index (RPI). The other is the consumer price index (CPI).

Input Output Analysis – This was used for countries, which have a system of national planning often called indicative planning. It tries to understand the inter-relationship between markets. It is helpful to have such a system if the government really wants to understand what relationships there are going to be for factors for production in the future which in turn will involve the transport system. It can also be used by individual firms.

Internal Rate of Return – This is found when using discounted cash flow (DCF); it means the rate of return on the capital employed.

Inventory Costs – Inventory costs are the cost of holding stocks. Traditionally many firms held considerable amounts of stocks not just for their own customers but also in the case of fuel for their role in the production process. Even commercial firms may hold stocks of spare parts in case the office machines have faults or need further parts. Sometimes firms may hold stocks where there are seasonal variations in price or availability for example stocks of potatoes are held so that they can be sold later in the year. Sometimes firms buying agricultural produce may if it is possible to store them wish to buy them when they are cheap in order to make a profit at a later stage. The same also applies to many other primary products.

Joint Costs – In transport, joint costs occur where one service can be used jointly for several different services. Joint costs could include for example track and signalling on former British Rail. One of the problems particularly of British Rail was that of allocating the joint costs in a sensible manner. At the present time Network Rail has the same problems.

Just in Case (JIC) – This is where a firm produces or holds goods on the assumption that customers might either physically see them or require them at any time.

Just in Time (JIT) – This is where a firm produces goods which are just in time for the given date of deliveries rather than the more conventional "just in case" when firms holds stock usually called inventories. This means that the firm concerned needs to know the ideal dates for not just manufacturing but also how long it will take to get the orders from its doors to that of the customers. Just in time caused problems during the fuel protests in 2001 when many lorries were unable to get enough fuel and so firms were unable to get their drivers to deliver the goods.

Lead Times – This is the period of time, which it takes between in some cases developing a project and then to bring it into fruition. It is also used in the freight sector to denote the time between placing an order and the time in which the firm can get its order to the customers. It will depend not just on the manufacturing process but also the transport firm concerned with the deliveries.

The lead time for the aircraft industry is often very considerable as the Dreamliner shows. The lead time for major projects such as the Channel Tunnel was considerable.

Load Factor – This can be defined as:

$$\frac{\text{Actual passenger kilometres or freight tonne kilometres}}{\text{Capacity}}$$

Load factors may be altered by changes in fares policy. Ideally, operators would like to have a 100% load factor since this will generally be the most profitable and make the best use of existing assets. Load factors on the railways and other operators are often low because of the problems of the peak.

Long Run Marginal Cost – This is defined as the additional cost of a unit of output when factors of production are not fixed. The long run marginal cost pricing policy would give the best criteria for investment given a number of limitations.

Macroeconomics – This deals with economic factors affecting the economy as a whole (e.g. unemployment, balance of payments and economic growth).

Marginal Cost – This is defined as the additional cost of a unit of output.

Merit Goods – These are goods or services which governments or the community may feel should be provided irrespective of the ability to pay. Examples of these in transport are less common than in health or education but might include ambulance services to and from hospital, provisions of basic transport services for poorer people including old age pensioners. This is possibly one of the reasons why people over 60 from 2008 in the UK have had free off-peak bus services. The age limit has been slightly altered over time.

Micro Economics – This deals with the theory of the firm: transport economics mainly deal with micro economics, although macro factors such as employment or unemployment may be important for firms to understand.

Mixed Economy – The majority of economies are mixed which means that they will include both public and private sector activities.

Monopoly – This means the sole producer of a good service for which there is no substitute in the eyes of the consumer. A complete monopoly is comparatively rare though from non-car owners in many rural areas a bus service may well be a monopoly. For long distance commuters to London on faster rail services the same may apply.

Monopsony – This is where there is only one buyer of a good or service. British Rail was a monopsony for signalmen. Whilst there are now many rail companies, in some parts of the country there is only one operator so this will still hold.

Multi Part Tariff – The use of a multi part tariff occurs when there is a significant difference between average and marginal costs. Multi part tariffs have several elements often a standing charge irrespective of usage and another charge for each unit of demand.

Multiplier – This is the relationship between any initial injection to the economy and the final effect on output often looked at in terms of gross national product (GNP), although economists may be interested in local multipliers, following the effect of transport investment into a transport project. In other cases, as with the EU, the EU may be interested in looking at the effects of bridging any obvious gaps in the transport infrastructure such as the fixed link between Sweden and Denmark. In the UK, the infrastructure commission headed by Lord Adonis a former Labour cabinet minister has been appointed to look at the multiplier effects amongst other ideas.

Net Present Value – This is used in investment appraisal. For this purpose, all future costs and benefits are discounted to the present day so that there is a common denominator.

Non-Price Competition – Where fares or charges are fixed whether because of collusion as in oligopoly or by government legislation firms are likely to indulge in non-price competition. This is partly because transport is not homogenous and some cases the market may be extended more as with the provision of new styles of comfort or through quicker services than by altering the price.

Obsolescence – This is where the commercial life of an asset is less than its technical life. This can either occur because of changes in demand for example, as in the case of many rural railways or because of changes in technology, for example, the last steam trains were built in the UK in 1960 with a technical life of at least 40 years, but were removed from the British Railways network by 1968.

Oligopoly – This is where there are only a few firms in an industry. The underlying economic problem of oligopoly is trying to determine price and output. There is no single theory, which would be able to predict firms' reactions in oligopoly. Oligopoly firms may decide to compete or to collude. They may compete on the basis of price or through non-price competition.

Operating Ratio – This is the ratio of working expenses to gross receipts as a percentage:

$$\frac{\text{Working expenses}}{\text{Gross receipts}} \times 100\%$$

Operating ratios can be used to compare different transport firms. It is sometimes used in the transport industry as a measure of efficiency, but this is not necessarily helpful since there may be a number of other factors for example how much of the firm's activity is carried out by outside contractors.

Opportunity Cost – This is one of the most important concepts in economics. It means what is given up. The economist's more precise definition is the cost of the next best alternative which has been foregone. Opportunity costs do not necessarily equal financial costs. For example, the opportunity cost of the Channel Tunnel before it was built is the vast amount of resources which were used. Once however it was built the opportunity cost is very small since there is little alternative use for it.

Pareto Optimum – This will occur if no one person can be made better off without another person being made worse off. Perfect competition will lead to the Pareto optimum assuming a number of stringent conditions.

Payback Period – This is a commonly used but is not very good method of measuring whether investment is worthwhile by estimating the time over which a project will pay for itself.

Perceived Costs – These are costs which customers or producers act on. There is considerable evidence to show that motorists' perceived costs are far smaller than their actual costs. One of the consequences of this is that demand for motoring measured in passenger kilometres is greater than if motorists acted in a purely rational way. The government could help to overcome this problem if it wished by raising the price of petrol. Transport operators competing with the car will need to take into account motorists' perceived costs.

Perfect Competition – This is a very important model which has influenced political thinking. It is the underlining concept in most of Keynesian economics and underlines much of monetarist economics. Tramp shipping is perhaps the only example in transport which approximates to this market model. The model has had a very important political influence since the Conservative Party in particular has often assumed that competition is preferable to monopoly power particularly by the public sector. Care however has to be taken when assuming that the public sector is necessarily monopolistic or that the private sector necessarily approximates to perfect competition.

Private Finance Initiative (PFI) – These have been widely used for major infrastructure investment where the private sector not just builds the infrastructure but also carries out maintenance. They have sometimes been seen as a method of reducing the public sector borrowing requirement (PSBR – it has now been renamed the public sector net cash requirement (PSNCR)) because of the odd method of classifying expenditure. They have come under attack particularly since Metronet went bankrupt leaving a great deal of expenditure still to be carried out on the Underground network in London. Jeremy Corbyn the current Labour leader in the UK has attacked the PFI concept.

Positive Economics – This is the part of economics where we can prove or disprove hypotheses with the use of data. We could say that cheaper fares offered by government to some people will have the effect of increasing travel by a certain amount and we can then look at data to see whether or not this has happened.

Predatory Pricing (Sometimes Known as Fighting to Kill) – This is where a firm has low charges for a short while in order to drive a rival out of business and then raises them again in that sector once it has eliminated the competition.

Price Discrimination – This is where different prices are charged which are unrelated to the different costs of provision. The railways for example have a large amount of price discrimination with their different fares for young people under 24, people over the age of 60 and the family card. Price discrimination may be used either for social purposes (e.g. in France there are special provisions for large families or for profitable reasons).

Price Elasticity of Demand – This is defined as:

$$\frac{\text{The percentage change in demand}}{\text{The percentage change in price}}$$

The figure is usually negative though the minus sign is often ignored. The economist uses the phrase inelastic if the figure is less than one, or elastic if the figure is greater than one. Demand for business travel is much more likely to be inelastic than demand for social travel. There may also be a distinction between the short run and long run price elasticity of demand since in the short run it is often more difficult to use a substitute transport service.

Private Costs – These are the costs paid by consumers or producers of transport services. The distinction between these and social costs is one of the reasons for the introduction of the cost benefit analysis technique.

Productivity – There are more problems in measuring productivity in transport than in many other industrial sectors. This is mainly because transport is not a homogeneous product. Productivity may be measured in terms of tonne kilometres or passenger kilometres per employee. On the freight side it could be measured by the number of tonnes handled per employee.

Production Possibility Frontier – This means the total amount of goods and services which a country or group of countries could produce. Unemployment which often affects the transport industry will reduce the total amount. Conversely, better training including apprenticeships will shift the production possibility frontier to the right (Figure 1).

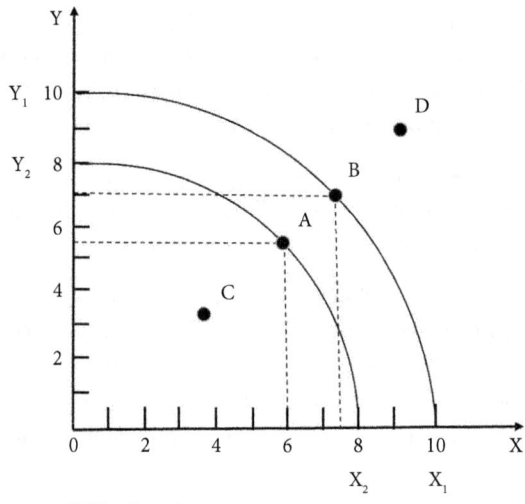

Figure 1 Production possibility frontier.

Profit – The accountant uses this term in a different way to that of the economist. For accounting purposes, profits vary according to whether a company is highly capital geared or has low capital gearing. It may also in a period of inflation depend upon the methods of depreciation and valuation of stocks.

The economists' definition includes imputed costs of labour and imputed costs of finance, whereas an accountant's definition will not.

Taxi Firm A		Taxi Firm B	
Trading Profit	£20,000	Trading Profit	£20,000
Managers Wage	£12,000	Owners imputed wage	£12,000
Finance Cost	£50,000 at 10% = £5,000	Owners capital	£50,000 at 10% = £5,000
Economists profit	£3,000	Economists profit	£3,000
Accounting Profit	£3,000	Accounting profit	£20,000

Profit Maximising – In conventional microeconomics, this has usually been assumed to be the major if not the only objective of a firm. In recent years, a variety of other objectives have also been suggested. These include alternative maximisation theories such as growth maximisation or revenue maximisation. Other economists have suggested that since we can rarely be sure that we have maximised anything that satisficing which means that firms set a series of targets for turnover, share of the market, level of profits and will be satisfied if they reach these targets is more plausible.

Progressive – The term progressive can apply to either taxation or subsidies. Taxation is said to be progressive if the rich pay proportionately more than the poor. Conversely, subsidies will tend to be progressive if they give more to the poor than the rich. Whilst in the UK taxation is usually progressive though the community charge (poll tax) was an exception to this, subsidies have not necessarily been progressive. For example, subsidies to Concorde or to company cars were almost certainly regressive.

Protectionism – Protectionism means where individual countries or sometimes groups of countries, keep out other countries' goods and services by imposing tariff barriers.

It has a substantial effect on the transport industry since if heavy taxes are imposed on foreign exports there will be less trade. It has become important for exporters to the USA with Donald Trump's ideas of keeping out foreign trade especially those which compete with the US car and other vehicles' manufacturers.

Public Corporations – These are bodies which are set up by a specific Acts of Parliament such as the 1947 or 1953 Acts in which the powers of the bodies and the economic objectives of these organisations are laid down.

Public Sector Borrowing Requirement (PSBR) – This is the gap between the amount of income the government receives and its expenditure. Partly because of monetarism the government has often made this a part of its macro-economic targets.

Public Service Obligation (PSO) – Under the Railways Act 1974 the railways had a public service obligation. They had to run a number of passenger services which were thought to be desirable but which would not have broken even.

Real Rate of Return – This means the rate of return after allowing for inflation so that if the rate of return was 8% in money terms but the rate of inflation was 3% then the real rate of return would be about 5%. We could find this more precisely by dividing 108 by 103 and then subtracting 100.

Real Terms – This means after allowing for inflation so that if we are looking at future revenue and cost we will usually use this since in many ways it is more realistic than just the money figure. This is especially true in period of high inflation rates as in the UK in the 1970s when the rate of inflation was at one stage around 25%.

Real Wages – This is the value of all income including fringe benefits, which arises from work particularly in the transport sector. It will often be higher than the nominal wage since it may include benefits such as future pensions as well as subsidised canteens or possibly company cars. It also is necessary to ensure that it is measured against inflation. A booking clerk in 1967 might have received a wage of around £1,000 a year compared with say £15,000 a year in 2009, but clearly, this does not indicate that a booking clerk would be 15 times as well off.

Regression Methods – This is a statistical term which is often used in transport economics to particularly look at forecasting of traffic demand. In its simplest form, it is likely to take the format that demand is a function of income and price.

Resource Cost – This means the opportunity cost of the resource used. It is not necessarily the financial cost since often there are taxes or subsidies. Sometimes the goods or service may have been already in existence as with railway track so that there is not a current financial cost. This may apply to road construction where prices are not generally charged directly.

Road Pricing – This term is used when car users pay directly towards the costs of congestion. In the UK, the Smeed Report recommended this in 1964. The idea is that marginal social costs should equal marginal private costs by charging motorists coming into the centre of town a congestion cost. Singapore has adopted a system of road pricing with exceptions for buses and cars carrying more than a certain number of passengers. Hong Kong, however, decided (before it reverted to Chinese rule) not to go ahead mainly because of political pressure. In London, the scheme started in February 2003 and the area was extended in February 2007.

Running Costs – These are the costs incurred in running a vehicle. For a car, this would include petrol or other fuel, oil, tyres and maintenance cost. It is approximately the same as the economists' concept of a variable cost. For an office, this could include fuel costs, telephone costs, printer cartridges, etc.

Satisficing – This is a term used by economists to indicate that in many cases firm may not be able to determine what profit maximising actually means, but instead they will set series of objectives such as an increase in the volume of traffic, profits, quality measures such as punctuality which give targets which can more readily be assessed as to whether or not they have been achieved.

Second Best – This term derives from welfare economics. It applies where "the best" policy is ruled out, for example road pricing as a means to alter the urban peak problem. Under such circumstances subsidies to public transport for the building of more transport infrastructure might be said to be a second best solution.

Sensitivity Testing – This concept is used when carrying out investment appraisals. It is extremely unlikely that forecasting of demand, prices, costs of maintenance, etc., can be completely correct. Therefore, it might be worthwhile testing a project by making a range of possible assumptions and seeing whether the project is still worthwhile.

Under such circumstances, the term "sensitivity testing" is used. In many cases, it is very difficult to forecast on the cost side items such as fuel as they may well be affected by political rather than straightforward economic considerations. Demand is noticeably difficult to forecast, particularly over long periods since it is often difficult to know what substitutes there may be for a mode of transport or even for transport as a whole. Therefore, economists using spreadsheets observe what will happen if the assumptions are varied.

The term is also used in a quite different way by the human relations department, to mean whether individuals and firms are sensitive to other internal and external stakeholders.

Shipping Conferences – These are ship-owner's organisations which gave a regular time-table service on major shipping routes. Customers who send all their goods to the particular destinations by members of the shipping conferences usually obtained a rebate at the end of the appropriate period.

Shipping conferences had the advantage of reducing fluctuations in prices, which might otherwise take place with perfect competition. Shipping conference operators also claimed that they gave a great more regular service and served a greater variety of destinations than with a free market.

Simulation – Often experiments are very expensive in transport, testing a new aircraft or making major changes to a transport network. Simulation is widely used for technical aspects since operators would not wish to test whether aircraft would be dangerous by using pilots before the project is underway and the same thing would apply to navigation. The use of computers is often used to simulate either physical characteristics testing out how much aircraft or ships might be subject to stresses and strains. Computers can also be used to show likely changes in demand if altering networks and timetables. Whilst simulations can be done without the use of computers, computers offer the ability to test a wide range of possibilities very quickly. The use of spreadsheets is common which can answer the questions quickly about the effects of changing some variables.

It is also now widely used for economic purposes operators might wish with the rail network to show what would happen if changes were made following the British exit from the EU.

Transport economists would suggest it should be used to look at the effects of electrification not just on one route but also elsewhere. If the line from St. Pancras which is already electrified as far as Bedford could be extended to Leicester or further to Derby or Nottingham this could obviously be useful in its own right, but also could well be useful as a diversionary route if there were problems on the East Coast Main Line or West Coast Main Line as there has been. Because transport is partly a matter of habit, returns might be greater than shown under conventional analysis since if passengers use other modes of transport where there are major projects going on, they may not return once they have been finished.

Social Costs – Social costs are total costs to the community and therefore include both private costs and external costs. The divergence of social and private costs has led to the use of cost benefit analysis being applied to many transport projects.

Standing Charges – This term is more frequently used in accounting than in economics. It approximates to the economist's concept of fixed costs. It generally is a term used for vehicles and means those charges which will occur even if no distance travelled. For example, with road haulage vehicles this will include vehicle excise duty, cost of garaging, cost of vehicle insurance and depreciation.

Supernormal Profits – These are profits over and above the normal rate of return. They cannot exist in perfect competition in the long term but can and do exist in either monopoly or oligopoly.

Supply Curve – The convention in economics is to show the quantity supplied on the horizontal axis and price to be shown on the vertical axis. The supply curve is usually upward sloping and shows how much producers will supply at a given price in any period of time.

Systems Analysis – Systems analysis whilst often being thought of as a technique mainly used by computer specialists can also be used within the transport industry without use of a computer. Often economists would first want to look at the existing data. This might be used initially sometimes this could be supplemented through the use of questionnaires or observations to find more information. The second step is to look at a viability study, for example whether a new transport system or altering even different types of vehicles such as those with higher speeds would be worthwhile. The third stage is that of designing the system, then of implementing the system and finally to tune the system sometimes known as tweaking.

Target Rates of Return – This was used in nationalised industries in the 1960s. Different industries were set different targets according to the amount of competition, etc. This was the rate of return which nationalised industries can make on total assets. In the 1960s they were often set different targets according to the type of competition.

Currently in the private sector, many organisations will set target rates of return partly so they can judge one part of the organisation against another. The vehicle assembly industry for cars will use this.

Tendering – This can occur in either the private or public sectors. It is used as for example where the local authority wants a network of services to be provided and where only one operator will usually be allowed to carry them out. To try to prevent collusion the tenders usually have to be placed in a plain envelope. The local authority or person asking for the tender does not always accept the lowest tender, since in many cases, the quality of service will ales be important. Tenders are also used quite frequently when tendering for a consultancy service. This has become increasingly common although some people have expressed concern about always going outside for consultancy rather than carrying out the work in-house.

Test Rate – This has often been used by the government in conjunction with the use of discounted cash flow techniques when considering investment. The test rate was sometimes expressed in real rather than in money terms. For example, in 1978 a White Paper suggested a minimum of 5% rate on new investment.

Theory of Games – This term is sometimes used in the study of oligopoly. The theory of games tries to show how firms may act taking into account other firms' likely reactions. Whilst the term is not always used many modern computer business games use the same approach, for example those from the British Institute of Management.

Currently Eurostar will be using the same approach to see how it can compete with the airlines and coaches for the railway routes from London to Paris and Brussels and also the planned extension to Amsterdam.

Time Valuation – The valuation of time is one of the important concepts in transport economics. It is particularly important in road investment since much of the benefits assumed from road building are savings in time. For road building unlike other projects there is usually no direct market and therefore economists have to find some way of valuing time.

Economists can try to look at the valuation of time by looking at the market. If there are two coach services travelling between the same town but one takes an hour longer and costs £5 more if passengers are willing to pay the £5 they value their time at least by £5 per hour. If they do not pay the £5 then the maximum valuation is £5. In practice it is often more complicated since it is comparatively rare for time to be the only difference between competing forms of transport. Comparing a coach and rail service between two towns the railway service may be more comfortable and have provision for businessmen to work such as computer terminals.

Even if we can find examples where time seems to be the only major difference would have to be to ensure that people are aware of the competing firm's differences in prices, time and quality of service. In London many people use the Underground services which are generally quicker than bus services. However, information about the Underground services is generally much greater than bus services and it would be unwise to base valuations on time on such observations. In practice people will often value small time savings at a different rate from longer time savings. Someone may be willing to pay £18 in order to save one hour on a journey, it does not follow that he or she will be willing to pay 30p for one-minute saving. Some economic studies noticeably Oxford University Transport Unit have tried to develop more advanced techniques such as HATS Household Activity Savings which observe people's overall changes in activities as a result of changes in transport. In practice as well research shows that people have a different value on leisure time to business time.

Total Distribution Costs – In freight transport transit costs i.e. the costs charged by transport companies are only part of the total distribution costs which would include insurance, costs of holding stocks, packaging, warehousing etc. Airline companies have often stressed this concept showing that their higher costs for carrying similar goods may well be outweighed by reductions in total distribution costs compared with other modes of transport.

Transparency of Subsidies – This is one of the features which many economists are strongly in favour of. Generally, with the EU competition is regarded as the best way to achieve economic welfare, but subsidies are allowed if they do not distort competition. Transparency of subsidies means that people can see what subsidies are being paid for. The EU as a whole would be concerned that subsidies were not paid to national carriers so that other goods and services are in effect being subsidised and thus distorting international competition.

Unit Loads – These include palletisation as well as container traffic. The road haulage vehicle itself is sometimes used as a unit, for example with the use of roll-on/roll-off ferries. The so-called unit load revolution or container revolution occurred in the 1960s.

Utility – This means what the consumer regards as helpful. It does not mean usefulness in the conventional sense. Some economists will criticise the obsession with fast speeds particularly of road vehicles at the expense of both fuel consumption and more importantly road accidents.

Value Judgements – These are statements, which cannot be proved or disproved through the use of scientific measurements. A great deal of transport economics is concerned with this. Government uses value judgements particularly in terms of public sector expenditure. For example, it is a value judgement to suggest that old age pensioners should be given cheap or free transport.

Value judgements will frequently apply with new or improved infrastructure where the government whether at central, local or supranational level has to decide whether benefits to the users outweigh the disadvantages to non-users.

Variable Costs – These are costs which vary with output. Examples of variable costs in transport would include fuel costs. In the short run transport firms will have to charge generally at least variable costs if they wish to stay in business.

Vertical Integration – This occurs where a transport firm takes over another firm which operates on a different level of activities such as an airline taking over a travel agent.

X Efficiency – This is a phrase publicised by Liebenstein in 1966. It refers to the idea that monopoly does not necessarily act in the efficient manner which most models had assumed and that instead firms need to be looked at to see how far efficient ways of production are being used if economic welfare is to be maximised.

Yield Management – This term has become more common particularly in the aviation industry. It refers to the idea that the vehicle should gain the maximum possible revenue from a particular journey. The easiest way to do this is generally through price discrimination. Yield management may become greater with better utilisation of vehicles. In 2017, however, Ryanair was heavily criticised because of the way it proposed to reduce the number of flights which had already been booked.

Index

A, B and C licensing system, 210
Accident rates, 198
Aggregated data, 21
AGR convention, 342–343
Airships, 191, 283
Alternatives to air transport, 182, 384
Alternatives to transport, 355
Arithmetic growth, 21
Armitage committee, 78, 376
Auto-wagon system, 218
Average costs, 61, 128, 318, 327, 387
Average fixed costs, 61
Average Rate of Return (ARR), 270

Baltic Exchange, 74
Beeching report (Reshaping of British railways), 8, 108, 128, 227, 284, 286, 312, 322
Berne Convention, 339
Boeing, 40, 166, 169, 182, 185–186, 279, 380
Break-Bulk Ships, 154
British Institute of Management Survey, 32
British Transport Commission (BTC), 215, 233
Buchanan report, 28, 148, 205, 233
Bucket shops, 80
Building vehicles within countries, 359
Bus bays, 123
Bus lanes, 10, 122, 253, 288, 310, 358
Bus roads, 123, 253, 309–310, 354, 388
Bus urban challenge fund, 125

Cabotage, 75
Capesize, 282
Car ownership and economic growth, 358
Car sharing, 22, 308, 382
Carnets, 96–97, 285, 305, 319, 343

Catamarans, 40, 91, 96, 158–159
Cellularships, 152
Central Business District (CBD), 11
Certificate of professional competence (CPC), 115
Chamber of Commerce, 155
Changes in location, 11, 31, 393
Chartered flights, 89, 100, 164, 191–192, 296
Chicago conference, 194, 342
Civil Aviation Authority, 75, 145, 342
Coach hours regulations, 68
Coarse fares, 95
Coastal services, 157
Collusion, 79–81, 90–91, 399, 407
Common external tariff (CET), 334
Common transport policy, 338, 395
Community partnerships, 318, 331
Company cars, 32, 213, 302, 310, 392, 403–404
Company trains, 217
Comparisons of costs, 261
Compensation principle, 238–239
Complementary goods or services, 390, 392
Concentration ratios, 83
Concessionary fares systems, 254
Concorde, 117, 141, 164, 176, 186–187, 216, 260, 336, 381, 403
Consumer surplus, 97–98, 237, 391
Co-ordination between transport and planning, 362
Countervailing power, 155
Cournot, 79
Cross elasticity of demand, 22–24, 31, 394
Cross section data, 247
Customer satisfaction, 264
Cycle tracks, 255, 352, 355

Daily peak, 69, 176, 295, 299
Decision making by the rich, 352
Demographic trends, 37
Demography, 27
Depreciation, 25, 64, 328, 392–393, 396, 402, 406
Deregulation, 229
Derived demand, 10, 33, 295, 393
Dial a bus services, 128
Discount rate, 272–273
Discounted cash flow, 212, 272, 274, 394, 397, 407

East London river crossing, 249
Enterprise zones, 256
Environmental Criticism of Airport Policy, 180
European airbus, 40, 169, 182, 185–187, 279, 336
Eurozone countries, 334

Fares policy, 302, 359, 398
Ferrovial, 229
Ferry service, 157
Fighting companies, 82
Fixed costs, 61–62, 106–107, 124, 129, 141, 318, 320, 328, 351, 354, 396, 406
Fixed exchange rates, 333
Flags of convenience, 156, 207, 261, 344
Flags of discrimination, 156
Flat rate, 96–98, 380
Food miles, 47
Footloose industry, 47
Forked tariff system, 82, 396
Foster Committee, 78
Freight Transport Association, 116, 279
Freight transport demand, 48
Freightliner, 108, 139, 147, 162, 225–227, 287, 340, 372

Geometric growth, 21
Gini coefficient, 349
Grand Central Railway, 76
Great Central line, 281, 372
Great North Eastern Railway (GNER), 76, 264, 288

Greater use of railways, 385
Greyhound services, 127
Guided bus ways, 124

Habit, 27–28, 51, 128–129, 140, 304, 367, 406
Hatfield disaster, 206
Heathrow congestion, 176
Heathrow Express, 8, 33, 172–173, 230, 299, 390
Herald of Free Enterprise, 197
Household Activity Travel Survey (HATS), 30
Hovercraft, 91, 146, 158, 226
Human Development Index (HDI), 348
Hydrogen cells, 29

Income elasticity of demand, 16, 18, 32, 54, 394
Independent Transport Commission, 214, 323
Inferior goods, 17, 335
Inland waterways, 145–146, 151, 153, 155, 157, 159–161, 232, 282
Interdependence, 7–8
Interdependent projects, 277
intermediate size vehicles, 353
Internal Rate of Return (IRR), 272, 275, 397
International Air Traffic Association (IATA), 190
International Civil Aviation Organisation (ICAO), 169–170, 194, 207
International collective agreements, 362
International Convention of the Taxation of Road Vehicles, 343
International Hydrographic Organisation, 207, 341
International Maritime Organisation (IMO), 52, 160, 207, 341
International Road Union, 344
Investment appraisal methods, 212, 267

Joint costs, 64–65, 211, 221, 227, 387, 397
Just in case (JIC), 52, 397
Just in time (JIT), 52, 251, 397

Kinked demand curve, 81
Kyoto, 5–6

Land use planning, 28, 56, 164, 243–245, 266, 314, 359, 382, 393
Law of unintentional consequences, 22
Lead time, 54, 259, 390, 398
Lighter aboard ship (LASH), 153
Limited stop services, 125, 304
Liner conferences, 154, 158, 362
Load factor, 49, 81, 100, 102, 107–108, 129, 180, 191, 193, 195, 212, 262–263, 310, 384, 398
Load line conventions, 207
Loading gauge, 50, 281, 284, 298, 340, 372, 389
London Docklands Development Corporation (LDDC), 33, 248
London Midland Scottish Railway (LMS), 2
London Passenger Transport Board, 251
London Transport, 87, 90, 97, 121, 123, 211, 242, 251–252, 259, 287, 303
Long run elasticities of demand, 14
Long run marginal costs, 62–63, 103–104
Los Angeles, 2

M1 motorway, 236
Maastricht treaty, 221
Marginal costs, 10, 62–63, 75, 103–105, 322, 387, 399
Market demand curve, 13
Matatus, 353–354
Minibuses, 326
Mobility management, 382
Monopoly Commission, 85, 88
MSC Napoli, 5
Multiplier effect of airports, 163

National Bus Company (NBC), 37, 86, 126, 229, 233, 252, 254, 326
National Freight Company, 221, 227, 233
National Freight Corporation, 211, 214, 221, 232–233, 289
National Ports Council, 210

Navigational aids, 140, 142, 144, 161, 186, 206, 341–342
Need for a low-cost public transport system, 352
Need for town planning, 354
Net Present Value (NPV), 272, 274–275, 399
Network South East card, 98, 106
New Towns, 33, 254–255
Noise problems, 165, 177
Normal goods, 17–18, 335

OBO ships, 153
October 2007 pre-budget report, 180, 384
Office of Fair Trading (OFT), 80
Office of Rail Regulation, 93, 228
Oligopoly, 78–82, 94, 387, 399, 406–407
OPEC, 4, 73, 81, 121, 247, 347–348
Open skies agreement 2007, 196
Open top buses, 129–130
Opportunity costs, 51, 65, 235, 400
Organisation for Economic Cooperation and Development (OECD), 347
Oyster cards, 96, 110, 301, 305, 385
Oyster smart card, 98

Panama Canal, 282
Parcel trains, 218
Pareto optimum, 74, 103–104, 209, 400
Passenger liners, 117, 158, 187, 191
Passenger Transport Executives, 84, 211, 242, 245, 252
Passenger travel combined with freight, 326
Payback period, 269–270, 400
Perceived costs, 25, 65, 75, 211, 322, 400
Perfect competition, 27, 73–75, 77, 78, 85, 87, 91, 94, 105, 209, 387, 400, 405–406
Population growth and difficulties in estimating population, 356
Port of London Authority (PLA), 166
Ports development, 351
Precision Approach Radar (PAR), 145
Predict and provide, 180, 244, 311
Prestige, 115, 117, 156, 341

Price discrimination, 94, 98–99, 104, 176, 263, 321, 401, 409
Price leadership, 80, 91
Private Finance Initiatives (PFI), 221
Productivity of staff, 264
Profitability, 61, 259–260, 270, 291–292
Proportion of fares bonus, 70
Pryke and Dodgson, 56, 69, 260
Public sector-borrowing requirement (PSBR), 221
Punctuality bonuses, 69
Purchasing Power Parity (PPP), 347

QE2, 288

Rail franchises, 374, 385
Rail Freight, 56, 65, 139, 217–219, 286, 339, 341, 368, 377, 380, 385
Railway drivers hours, 69
Railways and Transport Safety Act, 93
Rationality, 10, 23
Reclaiming derelict land, 249
Red Arrow, 304
Regional Spokes, 382
Restrictions on car ownership, 357
Road congestion, 23, 127, 160, 166, 237, 320, 357
Road pricing, 15, 209, 248, 260, 292, 302, 306–308, 310–311, 372, 377–378, 382, 404
Roll on/roll off, 57, 151–152, 330, 337, 351, 380, 383
ROSCOs, 113, 223, 390
Roskill Commission, 163, 165–169, 185, 238
Routeing of ships, 161
Rover tickets, 303, 329–330
Rural area peak, 296
Rural areas in the South East, 382
Ryanair, 193, 269, 286, 381, 409

Safety bonuses, 70
Safety stock, 54
Scottish bus group, 229

Seaplanes, 189
Season tickets, 30, 98–99, 110, 305, 374, 393
Seasonal peak, 175, 296
Sensitivity analysis, 276
Severance, 235, 237, 244
Sherman Act, 88
Short run elasticities of demand, 14
Short run marginal costs, 62–63, 103
Singapore, 15, 209, 302, 307, 312, 351, 358, 372, 404
Single market, 230, 334
SITPRO, 52, 344
Skewed income distribution, 348–349
Skilled workers, 349
Social travel, 14–15, 38, 100–101, 401
Sources of finance for airports, 174
Spare capacity, 102, 195, 210, 219
Split shift system, 69, 299, 306
Stacking areas, 142
Staff per vehicle or vehicle kilometre, 263
Standard gauge railways, 350
Standardisation of fleets, 195
Standby fares, 63, 102, 104
Status, 37, 228
Stern Report, 7
Stevenage, 254–255
Stock levels, 54–55
STOL, 37, 147, 163, 179, 189, 220, 226, 249, 343, 366, 368, 372
Structure plans, 245–246
Subsidies for public transport, 361
Subsidies for school usage, 323
Subsidies to aviation, 176
Subsidising cars, 324
Suez Canal, 282, 341
Supernormal profits, 73–74, 85, 406
Surveys of the industry, 257
Survival data, 263
Sweezy, 81

Tankers, 58, 88, 151, 153, 160, 166, 282, 341, 378
Tapering fares, 95–96

Test rate, 272–273, 407
Test report, 209
Thames Gateway area, 33
The Office of Rail Regulation, 93, 228
Time series data, 21
Tonnage, 45, 48, 57, 151, 280
Total distribution costs, 45, 338, 408
Trade creation, 334
Trade diversion, 334
Tramp shipping, 66–67, 74, 94, 154–155, 157, 400
Trans European Transport Network, 339
Transport and economic growth, 356
Transport for London, 98, 110, 125, 242, 252, 263
Transport in rural areas, 322, 355
Transport technology, 39, 50
Trinidad, 42
Trolleybuses, 33, 127
Turnover per employee, 265
Tyneside Metro system, 252

UK government review backs London Heathrow and London Stansted airport strategy, 179, 383
Ultra vires clause, 222, 228, 292
Union Internationale Chemin de Fer (UIC), 340

Variable costs, 3, 61–62, 73, 81–82, 106, 141, 153, 307, 322, 388, 409
Vertical integration, 263, 265, 289–290, 394, 409
VFR, 345
Victoria line, 110, 237
Visiting friends and relatives (VFR), 100, 345
Voluntary drivers for local services, 329
Voluntary railways, 20, 320–321
VTOL, 189

Walking bus, 41, 295
Watershed areas, 242
Wensleydale railway, 321
Willingness to pay principle, 238
Wilmott and Young, 16
Wood Enquiry in London, 48
Wood report, 279
Workable competition, 90
Wrexham Shropshire Marylebone Railway Company (WSMR), 293

Zero sum game, 80

Printed by BoD™in Norderstedt, Germany